Metal and Alloy Bonding: An Experimental Analysis

R. Saravanan · M. Prema Rani

Metal and Alloy Bonding: An Experimental Analysis

Charge Density in Metals and Alloys

 Springer

Dr. R. Saravanan
Research Centre and PG Department
  of Physics
The Madura College
Madurai 625 011
Tamil Nadu
India
e-mail: saragow@dataone.in;
       saragow@gmail.com

M. Prema Rani
Research Centre and PG Department
  of Physics
The Madura College
Madurai 625 011
Tamil Nadu
India
e-mail: premaakumar@yahoo.com

ISBN 978-1-4471-6178-3   ISBN 978-1-4471-2204-3 (eBook)
DOI 10.1007/978-1-4471-2204-3
Springer London Dordrecht Heidelberg New York

British Library Cataloguing in Publication Data
A catalogue record for this book is available from the British Library

Cover design: eStudio Calamar S.L.

Printed on acid-free paper

Springer is part of Springer Science+Business Media (www.springer.com)

Preface

This book has been written based on the experimental results obtained through several experimental techniques, especially the powder X-ray diffraction method, on various metals and alloys we encounter frequently. An analysis of the interactions of electrons in different atoms has been discussed.

Metals have useful properties including strength, ductility, high-melting points, thermal and electrical conductivity and toughness. The key feature that distinguishes metals from non-metals is their bonding. The existence of free electrons in metals has a number of profound consequences for the properties of metallic materials. There are a large number of possible combinations of different metals and each has its own specific set of properties. The physical properties of an alloy, such as density, reactivity, Young's modulus and electrical and thermal conductivity, may not differ greatly from those of its elements, but engineering properties, such as tensile strength and shear strength, may be substantially different from those of the constituent materials. Metals and their alloys make today's manufacturing industry, agriculture, construction and communication systems, transportation, defense equipments, etc. possible. Some of the major reasons for the continuing advancements in alloys are the availability of materials, new manufacturing techniques and the ability to test alloys before they are produced. Most modern alloys are, in fact, preplanned using sophisticated computer simulations, which help to determine what properties they will display. Semiconductors have been studied extensively due to their importance in applications. These materials receive much attention because physical properties such as the band gap, mobility and lattice parameter can be continuously controlled. Having such continuous control is of importance in applications such as electronic and optical devices.

Metals and alloys have high-melting temperatures because of the heavy bonding between the atoms. There are a variety of applications for metals and alloys. Due to the importance of these materials, a study of their bonding interactions has been carried out in this monograph using experimentally observed X-ray diffraction data.

Today's technological evolution results in developing new and sophisticated materials of immense use in domestic, technical and industrial applications.

Usually, the synthesis of new materials, especially metals and alloys, results in single-phase materials, but often not in single crystalline form. Hence, a complete analysis of the structure, local distribution of atoms and electron distribution in core, valence and bonding region is necessary using powder diffraction methods, in addition to single crystal diffraction results, since most of the recent materials will be initially obtained in powder form. Since one can make efforts to grow single-crystals from powders, a prior analysis is required using powders to proceed for single crystal growth.

In this context, we have taken some simple metals (Al, Cu, Fe, Mg, Na, Ni, Te, Ti, Sn, V, Zn) and alloys (AlFe, CoAl, FeNi, NiAl) and collected powder XRD data sets or used single crystal XRD data sets from the literature, to study the structure in terms of the local and average structural properties using pair distribution function (hereafter PDF), electron density distribution between atoms using Maximum Entropy Method (hereafter MEM) and bonding of core and valence electron distribution using multipole technique. Particularly, the PDF analysis requires data sets of very high values of Q ($=4\pi Sin\theta/\lambda$) which is achievable only through synchrotron studies, but not accessible for common crystallographer/material scientists. The present work gives reasonable results obtained through single crystal work or through high Q data sets, using only powder samples. Also, a study on the electronic structure of metals using the most versatile currently available techniques like MEM and multipole method is worthwhile. If the tools available for analysis yield highly precise information, then it is appropriate to apply it to precise data sets available as in this work, and thus the methodology can also be tested. In order to elucidate the distribution of valence electrons and the contraction/expansion of atomic shells, multipole analysis of the electron densities was also carried out. Recently, multipole analysis of the charge densities and bonding has been widely used to study the electronic structure of materials.

Bonding studies in crystalline materials are very important, especially in metals, because of their extensive use. These studies can reveal the qualitative nature of bonding as well as the numerical values of mid-bond densities which indicate the strength of the material under study. With the advent of versatile methods like MEM and multipole method, bonding studies gained impetus because of the accuracy of these methods and the fact that the experimental data can be used with these methods to accurately determine the actual bonding between atoms.

The precise study of bonding in materials is always useful and interesting, yet no study can reveal the real picture as no two sets of experimental data are identical. This problem is enhanced when the model used for the evaluation of electron densities is not entirely suitable. Fourier synthesis of electron densities can be of use in picturing bonding between two atoms, but it suffers from the major disadvantages of series termination error and negative electron densities which prevent the clear understanding of bonding between atoms; the factor intended to be analysed. The advent of MEM solves many of these problems. MEM electron densities are always positive and even with limited number of data, one can determine reliable electron densities resembling true densities. Currently, the multipole analysis of charge densities has been widely used to study crystalline materials. This synthesizes the electron density of an atom into core and valence parts and yields an accurate picture of bonding in a crystalline system.

In this research monograph on metals and alloys, a complete analysis of bonding has been made on 11 important metals and four alloys. Powder X-ray diffraction data as well as single crystal data sets have been used for the purpose.

Charge density analysis of materials provides a firm basis for the evaluation of the properties of the materials. Designing and engineering of new combinations of metals requires firm knowledge of the intermolecular features. Recent advances in technology and high-speed computation has put the crystal X-ray diffraction technique on a firm pedestal as a unique tool for the determination of charge density distribution in molecular crystals. Methods have been developed to make experimental probes to unravel the features of charge densities in the intra and intermolecular regions in crystal structures. In this report the structural details have been elucidated from the X-ray diffraction technique through Rietveld technique. The charge density analysis has been carried out with MEM and multipole method, and the local and average structure analysis by atomic PDF.

This research work reveals the local and average structural properties of some technologically important materials, which are not studied along these lines. New understandings of the existing materials have been gained in terms of the local and average structures of the materials. The electron density, bonding, and charge transfer studies analysed in this work will give fruitful information to researchers in the fields of physics, chemistry, materials science, metallurgy, etc. These properties can be properly utilized for the proper engineering of these technologically important materials.

Chapter 1 introduces the significance and applications of metals, alloys and semiconductors studied in this research work. The objectives of this book are presented. The essential mechanism of ball milling which has evolved to be a simple and useful method for the formation of nano crystalline materials is discussed. The current state of art of non-destructive characterisation techniques such as X-ray diffraction and scanning electron microscope are discussed.

Chapter 2 provides a survey of the current applications of X-ray diffraction techniques in crystal structure analysis, with focus on the recent advances made in the scope and potential for carrying out crystal structure determination directly from diffraction data. The basic concepts of crystal structure analysis, Rietveld refinement and the concepts used for the estimation and analysis of charge density in a crystal are discussed. The more reliable models for charge density estimation like multipole formalism and MEM are discussed in detail. The local structural analysis technique and atomic PDF is also discussed.

Chapter 3 presents the results and discussions of this research work. A detailed account of the results of the materials analysed are presented in the subsections.

Section 3.1 (Sodium and Vanadium Metals) describes about the nature of bonding and the charge distribution in sodium and vanadium metals are analysed using the reported X-ray data of these metals. MEM and multipole analysis used for bonding in these metals are elucidated and analysed. The mid-bond densities in sodium and vanadium are found to be 0.014 and 0.723 e/Å^3 respectively, giving an indication of the strength of the bonds in these materials. From multipole analysis, the sodium atom is found to contract more than the vanadium atom.

Section 3.2 (Aluminium, Nickel and Copper) describes the average and local structures of simple metals Al, Ni and Cu are elucidated for the first time using MEM, multipole and PDF. The bonding between constituent atoms in all the above systems is found to be well pronounced and clearly seen from the electron density maps. The MEM maps of all the three systems show the spherical core nature of atoms. The mid-bond electron density profiles of Al, Ni and Cu reveal the metallic bonding nature. The local structure using PDF profile of Ni has been compared with that of the reported results. The R value in this work using low Q XRD data for the PDF analysis of Ni is close to the value reported using high Q synchrotron data. The cell parameters and displacement parameters were also studied and compared with the reported values.

Section 3.3 (Magnesium, Titanium, Iron, Zinc, Tin and Tellurium) describes the average and local structures of magnesium, titanium, iron, zinc, tin and tellurium are analysed using the MEM, and PDF. The structural parameters of the metals were refined with the well-known Rietveld powder profile fitting methodology. One-, two- and three-dimensional electron density distributions of Mg, Ti, Fe, Zn, Sn and Te have been mapped using the MEM electron density values obtained through refinements. The mid-bond density in Ti is the largest value along [110] direction among the six metal systems. From PDF analysis the first neighbour distance is observed to decrease as the atomic number increases for all the metals.

Section 3.4 (Cobalt Aluminium and Nickel Aluminium Metal Alloys) describes the precise electron density distribution and bonding in metal alloys CoAl and NiAl is characterized using MEM and multipole method. Reported X-ray single-crystal data used for this purpose. Clear evidence of the metal bonding between the constituent atoms in these two systems is obtained. The mid-bond electron densities in these systems are found to be 0.358 and 0.251 e/Å^3 respectively, for CoAl and NiAl in the MEM analysis. The two-dimensional maps and one-dimensional electron density profiles have been constructed and analysed. The thermal vibration of the individual atoms Co, Ni and Al has also been studied and reported. The contraction of atoms in CoAl and expansion of Ni and contraction of Al atom in NiAl is found from multipole analysis, in line with the MEM electron density distribution.

Section 3.5 (Nickel Chromium ($Ni_{80}Cr_{20}$)) describes the alloy $Ni_{80}Cr_{20}$ was annealed and ball milled to study the effect of thermal and mechanical treatments on the local structure and the electron density distribution. The electron density between the atoms was studied by MEM and the local structure using PDF. The electron density is found to be high for ball-milled sample along the bonding direction. The particle sizes of the differently treated samples were realized by SEM and through XRD. Clear evidence of the effect of ball milling is observed on the local structure and electron densities.

Section 3.6 (Silver doped in NaCl ($Na_{1-x}Ag_xCl$)) describes the alkali halide $Na_{1-x}Ag_xCl$, with two different compositions ($x = 0.03$ and 0.10) is studied with regard to the Ag impurities in terms of bonding and electron density distribution. X-ray single crystal data sets have been used for this purpose. The analysis focuses on the electron density distribution and hence the interaction between the atoms is clearly revealed by MEM and multipole analysis. The bonding in these systems is studied using two-dimensional MEM electron density maps on the (100) and (110) planes and one-dimensional electron density profiles along the [100], [110] and [111] directions. The mid-bond electron densities between atoms in these systems are found to be 0.175 and 0.183 e/Å^3, respectively, for $Na_{0.97}Ag_{0.03}Cl$ and $Na_{0.90}Ag_{0.10}Cl$. Multipole analysis of the structure is performed for these two systems, with respect to the expansion/contraction of the ion involved.

Section 3.7 (Aluminium Doped with Dilute Amounts of Iron Impurities (0.215 and 0.304 *wt% Fe*)) describes the electronic structure of pure and doped aluminium with dilute amounts of iron impurities (0.215 and 0.304 wt % Fe) has been analysed using reported X-ray data sets and the MEM. Qualitative as well as quantitative assessment of the electron density distribution in these samples is made. The mid-bond characterization leads to a conclusion about the nature of doping of impurities. An expansion of the size of the host aluminium atom was observed with Fe impurities.

Chapter 4 presents the conclusion of the results of the reported work.

A complete analysis on the electron density of important metals and alloys is presented in this book. This book will be highly useful for scientists and researchers working in the areas of metallurgy, materials science, crystallography, chemistry and physics.

Acknowledgments

The author Dr. R. Saravanan, acknowledges his family for their kind support, help and for making the atmosphere conducive during the course of the compilation of this book.

The author Ms. M. Prema Rani, wishes to thank her family, husband and especially her children for their support and for motivating her in writing this book.

The authors thank the various finding agencies in India, the University Grants Commission (UGC), Council of Scientific and Industrial Research (CSIR) and Department of Science and Technology (DST), though they did not fund the compilation of this book directly. But, the authors believe that the various research tasks accomplished during the course of the work for the book may involve usage of the resources arising out of the funds by the above agencies and hence these agencies are gratefully acknowledged.

The authors wish to render their cordial thank to the authorities of the Madura College, Madurai, 625 011, India for their generous support in the various research efforts by the authors which led to the successful compilation of this book. Research of high quality needs good support from various people including the authorities in the concerned institutions from where the research efforts originate. In that respect, the authors thank the principal and the board of management of the Madura College, Madurai, 625 011, India, particularly the secretary, Mr. M.S. Meenakshi Sundaram, The Madura College Board, Madurai, 625 011, India for his support and encouragement in the academic and research efforts of the authors.

Editing a book on a special topic like the present one involves help, support, and constant motivation by a large number of clause of people, right from clerical level and up to intellectual level. The authors wish to acknowledge all those people who could not find a place in this page of this book but who rendered their cordial help for successfully editing this book.

The authors dedicate this book for real hard working people with real positive qualities.

Dr. R. Saravanan
M. Prema Rani

Contents

Chapter 1
Introduction

Abstract The properties of a material are a direct result of its internal structure. The ability to control structures through processing, and to develop new structures through various techniques, requires qualitative and quantitative analysis of the atomic and electronic structure. The average and local structure of some significant metals and important alloys have been analyzed and reported in this book. This introduction chapter deals with the significance and applications of metals, alloys and semiconductors. The essential mechanism of ball milling which has evolved to be a simple and useful method for the formation of nano crystalline materials is discussed. The current state-of-the-art of non-destructive characterisation techniques such as X-ray diffraction and scanning electron microscope are discussed.

1.1 Introduction

The development of improved metallic materials is a vital activity at the leading edge of science and technology. Metals offer various combinations of properties and reliability at a cost which is affordable. They are versatile because subtle changes in their microstructure can cause dramatic variations in their properties. An understanding of the development of microstructure in metals, rooted in thermodynamics, crystallography and kinetic phenomena is essential for the materials scientist.

Alloys can blend the properties of two or more metals to create a hybrid metal that is more cost-effective, stronger, more durable and overall better suited to its intended purpose than the pure metals used to create the compound. With emerging requirement of designing new materials capable of sustaining high-strain rate and severe operating conditions with reduced wastage of cost, energy and material, it has become an important issue to develop full understanding of

R. Saravanan and M. Prema Rani, *Metal and Alloy Bonding: An Experimental Analysis*,
DOI: 10.1007/978-1-4471-2204-3_1, © Springer-Verlag London Limited 2012

the nature of enhanced mechanical properties of the materials. New materials that can be tailored for individual applications are always in a constant demand. As the range of uses for powder metallurgy, hard metals and electronic materials expands, customer requirements are causing materials companies to come up with new products that have the required properties.

1.2 Significance of the Present Work

Metals and semiconductors play an important role in the present world as evidenced by their variety of applications. Hence, a study on some important metals, alloys and semiconducting systems is essential in terms of the local structure and the average structure which are completely different. The usual methods of analysis using structural refinement of X-ray or neutron data will give only the average structure of the materials under investigation. The studies on the local structure of materials seem to be rare because of the complexity of the problem. There is only limited information available about the investigations of materials in terms of the local structure. Numerous research papers are being published every year based on powder as well as single-crystal X-ray diffraction (XRD) data. The structures reported using those data are only average structures. Since, the analysis of local structure requires highly precise data up to maximum possible Bragg angle, accurate refinement of the data is limited. Due to the complexity of the problem, tasks of acquirement of precise X-ray data from the samples, and the computational incapabilities, local and average structural analysis has not been much explored. Atomic ordering is closely related to the materials' electronic and magnetic properties. Although the physical properties of alloys are closely related to their electronic structures, studies on the charge transfer and hybridisation of the electronic states are still insufficient (Lee et al. 2004).

In the present monograph, apart from pure metals, investigations on the local and average structures of doped metals and alloys are carried out with various doping concentrations. The average structure has been studied using both single-crystal and powder XRD data in some cases. The bonding and electron density distribution of the host as well as dopant atoms have been studied using tools like maximum entropy method (MEM) (Collins 1982) and multipole analysis (Hansen and Coppens 1978). For powder analysis, Rietveld refinement technique (Rietveld 1969) (for average structure) and Pair Distribution Function (Proffen and Billinge 1999) (for local structure) have been used. Effects on the electron density distribution by ball milling (El-Eskandarany 2001; Suryanarayana 2004; Ares et al. 2005) of alloy has been analyzed in this work.

The present research work reveals the local and average structural properties of some technologically important materials, which are not studied in these lines. New understandings of the existing materials have been gained in terms of the local structure and average structure of the materials. The electron density, bonding and charge transfer studies analyzed in this work would give fruitful information to researchers in the fields of physics, chemistry, materials science, metallurgy, etc. These properties can be properly utilised for the proper engineering of these technologically important materials.

1.3 Objectives

Though the materials studied and reported in this research book are all metals and alloys, the work has been divided into several parts for the sake of convenience. They have been given as below.

1. The average and electronic structure of the following elemental metals using Rietveld (Rietveld 1969), multipole (Hansen and Coppens 1978) and MEM (Collins 1982) by single-crystal XRD data.

   - Sodium (Na)
   - Vanadium (V)

2. a. The local, average and electronic structure of the following elemental metals using Rietveld (1969), and MEM (Collins 1982) by powder XRD data.

   - Magnesium (Mg)
   - Aluminium (Al)
   - Titanium (Ti)
   - Iron (Fe)
   - Nickel (Ni)
   - Copper (Cu)
   - Zinc (Zn)
   - Tin (Sn)
   - Tellurium (Te)

   b. The local structural information by analyzing the atomic pair distribution function (PDF) (Proffen and Billinge 1999).

3. The average and electronic structure of the following metal alloys using multipole (Hansen and Coppens 1978) and MEM (Collins 1982) by single-crystal XRD data.

   - cobalt aluminium (CoAl)
   - nickel aluminium (NiAl)
   - Iron nickel (FeNi)

4. a. The annealing and ball milling of the alloy nickel chromium ($Ni_{80}Cr_{20}$)

   b. The study of the local, average and electronic structure of the annealed and ball milled alloy $Ni_{80}Cr_{20}$ using Rietveld (Rietveld 1969) and MEM (Collins 1982) by powder XRD data.

   c. To study the local structure using PDF (Proffen and Billinge 1999).

   d. Analysis of the particle sizes of the differently treated samples by scanning
      electron microscopy (SEM) and (XRD).
5. The study of the average and electronic structure of the following doped alloys
   using Rietveld (1969), multipole (Hansen and Coppens 1978) and MEM
   (Collins 1982) by single-crystal XRD data.

   - Sodium chloride with iron impurities ($Na_{1-x}Ag_xCl$)
   - Aluminium, with iron impurities (0.215 wt% Fe and 0.304 wt% Fe)

1.4 Metals

Metals account for about two-thirds of all the elements and about 24% of the mass
of the planet. Metals have useful properties including strength, ductility, high-
melting points, thermal and electrical conductivity and toughness. The key feature
that distinguishes metals from nonmetals is their bonding. (Gallagher and Ingram
2001) Metallic materials have free electrons that are free to move easily from one
atom to the next. The existence of these free electrons has a number of profound
consequences in the properties of metallic materials (Kittel 2007).

The local and average structures of some technologically important metals such
as sodium, magnesium, aluminium, titanium, vanadium, iron, nickel, copper, zinc,
tin and tellurium are analyzed in this work and their typical properties and uses are
presented below.

1.4.1 Sodium

Sodium is a soft, silvery-white, highly reactive metal having only one stable isotope;
^{23}Na. Sodium ion is soluble in water in nearly all of its compounds. Sodium metal is
so soft that it can be cut with a knife at room temperature (Zumdahl 2007)

Sodium compounds are important for the chemical, glass, metal, paper,
petroleum, soap, and textile industries. A sodium–sulphur battery is a type of
molten metal battery constructed from sodium and sulphur. This type of battery
has a high-energy density, high efficiency of charge/discharge (89–92%) and long
cycle life, and is fabricated from inexpensive materials (Oshima et al. 2004).
NaS batteries are a possible energy storage technology to support renewable
energy generation, specifically in wind farms and solar generation plants. In the
case of a wind farm, the battery would store energy during times of high wind but
low-power demand. This stored energy could then be discharged from the batteries
during peak load periods. In addition to this power shifting, it is likely that sodium
sulfur batteries could be used throughout the day to assist in stabilising the power
output of the wind farm during wind fluctuations (Walawalkar et al. 2007). Due to
its high-energy density, the NaS battery has been proposed for space applications
(Auxer 1986).

1.4.2 Vanadium

Pure vanadium is a bright white metal, and is soft and ductile. It has good corrosion resistance to alkalis, sulphuric and hydrochloric acid and salt water. The metal has good structural strength and a low fission neutron cross section, making it useful in nuclear applications. (Lynch 1974).

Vanadium is used in producing rust resistant, spring and high-speed tool steels. It is an important carbide stabiliser in making steels. Vanadium is also used in producing superconductive magnets with a field of 175,000 gauss (Lide 1999).

The role of vanadium complexes in catalytically conducted redox reactions (Crans et al. 2004) and potential medicinal applications, such as in the treatment of diabetes type I and type II (Crans 2000), has stimulated interest in the stereo-chemistry and reactivity of its coordination compounds (Monfared et al. 2010).

Vanadium oxides and vanadium oxide-related compounds have a wide range of practical applications such as catalysts, gas sensors and cathode materials for reversible lithium batteries, electrochemical and optical devices, due to their structural, novel electronic and optical properties (Zhang et al. 2010). Mixed metal oxides find applications in a variety of fields due to the wide variation in their dielectric and electrical properties. The vanadium-based oxide ceramics have high-dielectric constant, low-dissipation factor and high-quality factor, which favour the use of these ceramics in many fields (Nithya and Kalaiselvan 2011).

Secure and reliable power is essential in areas such as telecommunications and information technology to safeguard the vast computer networks that have been established. Uninterruptible power systems have incorporated battery technology to allow smooth power feeding switch-over in the case of a power failure. In such systems lead-acid batteries are commonly being used until generators come online or for safe computer shutdown. The vanadium redox battery provides many advantages over conventional batteries for emergency back-up applications. This system stores all energy in the form of liquid electrolytes which are re-circulated around the battery system. The electrolytes can be recharged for indefinite number of times, or the system can be instantly recharged by mechanically exchanging the discharged solution with recharged solution (Kazacos and Menictas 1997).

Figure 1.1 shows a vanadium redox battery. A vanadium redox battery consists of a power cell in which two electrolytes are kept separated by an ion exchange membrane. Both the electrolytes are vanadium based. Vanadium redox batteries are based on the ability of vanadium to exist in four different oxidation states (V_2, V_3, V_4 and V_5), each of which holds a different electrical charge. The electrolyte in the negative half-cell has V_3^+ and V_2^+ ions, while the electrolyte in the positive half-cell contains V_3^+ and V_2^+ ions. During charging, reduction in the negative half-cell converts the V_3^+ ions into V_2^+ ions. During discharge, the process is reversed, oxidation in the negative half-cell converts V_2^+ ions back to V_3^+ ions. The typical open-circuit voltage created during discharge is 1.30 V at 25°C (Skyllas-Kazacos 2003).

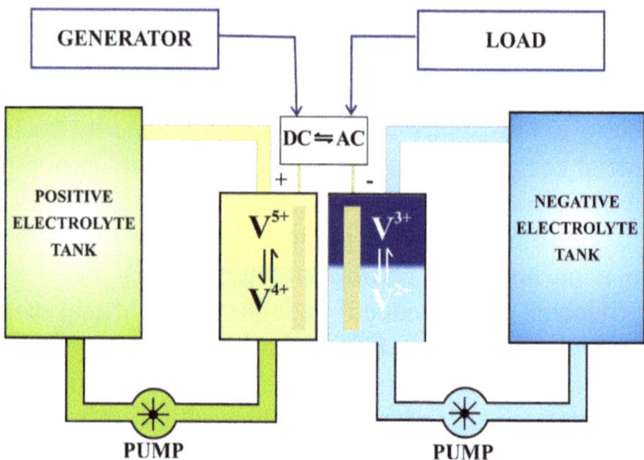

Fig. 1.1 Vanadium redox battery

Other useful properties of Vanadium flow batteries are their quick response to changing loads and their extremely large overload capacities. Their extremely rapid response times also make them perfectly well suited for UPS-type applications, where they can be used to replace lead-acid batteries and even diesel generators.

1.4.3 Magnesium

Elemental magnesium is a fairly strong, silvery-white, light-weight metal. The lightness combined with good strength-to-weight ratio has made magnesium and its alloys suitable for use in missiles and automotive industry.

Magnesium alloys have low density (1.5–1.8 g/cm^3) and high strength in relation to their weight (Kainer 2000). Magnesium alloys are used for die-casting due to their good corrosion resistance and low heat of fusion with the mould material. Most of the magnesium alloy castings are made for the automotive industry. Lowering car weight by 100 kg makes it possible to save 0.5 l petrol/ 100 km. It is anticipated that in the following years the mass of castings from magnesium alloys in an average car will rise to 40 kg, internal combustion engines will be made mostly from the magnesium alloys and car weight will decrease from 1,200 to 900 kg (Mordike and Ebert 2001a, b).

A good capability of damping vibrations and low inertia connected with a relatively low weight of elements have predominantly contributed to the employment of magnesium alloys for the fast moving elements and in locations where rapid velocity changes occur; some good examples may be car wheels, combustion engine pistons, high-speed machine tools and aircraft equipment elements (Wang et al. 2002).

The concrete examples for the use of castings of magnesium alloys in batch production in the automotive industry are elements of the suspension of the front and rear axes of cars, propeller shaft tunnel, pedals, dashboards, elements of seats, steering wheels, elements of timer-distributors, air filters, wheel bands, oil sumps, elements and housings of the gearbox, framing of doors and sunroofs and others (Dobrzánski et al. 2007).

In recent times, the increased environmental concerns and the rising costs of oil have again made magnesium and its alloys a material of interest for the automotive industry. Considering the characteristics of low density of magnesium, its extensive use in structural body parts of vehicles will offer major reductions of weight and hence reduction in fuel consumption. Such weight reduction provides a significant contribution to reducing the carbondioxide emission. It is estimated that an average new car produces 156 g CO_2/km travelled. This could be reduced to around 70 g CO_2/km through the application of magnesium technology (Mehta et al. 2004).

The advantages of magnesium and magnesium alloys are, lowest density of all metallic constructional materials, high-specific strength, good castability, suitable for high-pressure die-casting, can be turned or milled at high speed, good weldability under controlled atmosphere, much improved corrosion resistance, readily available, better mechanical properties, resistant to ageing, better electrical and thermal conductivity and recyclability (Mordike and Ebert 2001a, b).

Magnesium alloys have attracted increasing interest in the past few years due to their potential as implant materials. Magnesium and its alloys are degradable during their time of service in the human body. Magnesium alloys offer a property profile that is very close or even similar to that of human bone (Hort et al. 2010).

1.4.4 Aluminium

Aluminium has been the dominant material in the aircraft industry for more than a half century due to its attractive combination of light weight, strength, ductility, corrosion resistance, ease of assembly and low cost (Dorward and Pritchett 1988). Aluminium foam sandwiches (AFS) due to their flexible process ability and potential of cost reduction find application in space components. Currently, these light-weight materials find some first applications in particular fields of mechanical engineering such as race cars, and small series of other land-based vehicles (Schwingel et al. 2007). The use of high-strength aluminium alloys in automotive and aircraft industries allows reducing significantly the weight of the engineering constructions. In these fields, very often the main requirements for the components include high fatigue and wear-resistance (Lonyuk et al. 2007). Aluminium solar mirrors are an alternative for solar concentrators. The aluminium reflectors often offer an initial reflectance of 85–91% for solar irradiance. They have good mechanical properties and are easy to recycle (Almanza et al. 2009). The high strength-to-weight advantage of aluminium alloys has made it the material of choice for building airplanes and sometimes for the construction of land-based structures. For marine applications, the use of high-strength, weldable and corrosion-resistant aluminium alloys have made it the material of choice for weight sensitive applications such as fast ferries, military patrol craft and luxury yachts and to lighten the top-sides of offshore structures and cruise ships (Paik et al. 2005).

1.4.5 Titanium

Titanium has many desirable physical properties. The pure metal is relatively soft and weak, but it becomes much stronger when mixed with other metals to form alloys. The high-melting point of titanium (1,668°C) shows that it is an ideal material for the construction of high-speed aircraft and space vehicles.

Due to the exceptional strength-to-weight ratios, toughness, high stiffness and excellent biocompatibility, titanium and its alloys are used extensively in aerospace, chemical and biomedical applications (Kartal et al. 2010). Titanium alloys are widely used in the aerospace industry due to their excellent fatigue/crack propagation behaviour, and corrosion resistance (Markovsky and Semiatin 2010). The alloys of titanium represent significant advantages over most other engineering materials used for a variety of industrial applications due to their resistance to corrosion, oxidation and erosion. Titanium and its alloys have high-chemical durability as well as high strength. Their use is significant in nuclear industry, since the mechanical strength, high-heat proof and radiation proof are desired by many components, such as the steam condenser tubes, the irradiation targets for transmuting radioactive wastes and the overpacks for geological disposal of high level radioactive wastes (Setoyama et al. 2004).

Titanium is the preferred choice for surgical instrumentation due to its lighter weight, bacterial resistance and durability. High strength-to-weight ratio, corrosion resistance, non-toxic state and non-ferromagnetic property has made titanium "the metal of choice" within the field of medicine. It is also durable and long-lasting.

When titanium cages, rods, plates and pins are inserted into the body, they can last for more than 15 years. And dental titanium, such as titanium posts and implants, can last even longer. Osseo integration is a unique phenomenon where the body's natural bone and tissue actually bonds to the artificial implant. This firmly anchors the titanium dental or medical implant into place. Titanium is the only metal that allows this integration. Titanium and its alloys are widely used to replace failed hard tissues, such as artificial hip joints and dental implants (Li et al. 2008). Mechanical properties such as high strength, ductility and fatigue resistance, as well as a low modulus make titanium and its alloy suitable for applications in jet propulsion systems and human body implant (Heinrich et al. 1996). Titanium has long been used as an implant material in different medical applications, showing excellent performance in forming a close contact to the surrounding tissues (Petersson et al. 2009).

1.4.6 Iron

Pure iron is silvery metal with a shining surface. It is a good conductor of heat and electricity. Iron is used to make bridges, automobiles and support for buildings, machines and tools. It is mixed with other elements to make alloys, the most important of which is steel (Sparrow 1999).

Iron-based glassy alloys seem to be one of the most interesting materials due to their soft magnetic properties including high-saturation magnetisation. They are suitable materials for many electrical devices such as electronic measuring and surveillance systems, magnetic wires, sensors, band-pass filters, magnetic shielding, energy-saving electric power transformers (Nowosielski et al. 2008).

1.4.7 Nickel

Nickel is a silvery-white lustrous metal with a slight golden tinge. It is one of the four ferromagnetic elements that exist around room temperature, the other three being iron, cobalt, and gadolinium. Its Curie temperature is 355°C. Nickel is non-magnetic above this temperature (Kittel 1996). Nickel belongs to the transition metals and is hard and ductile. The isotopes of nickel range from ^{48}Ni to ^{78}Ni. The isotope of nickel with 28 protons and 20 neutrons ^{48}Ni is "double magic" and therefore unusually stable (Audi 2003).

The metal is corrosion-resistant, finding many uses in alloys, as plating, in the manufacture of coins, magnets, common household utensils, rechargeable batteries, electric guitar strings, as a catalyst for hydrogenation, and in a variety of other applications. Enzymes of certain life-forms contain nickel as an active centre, which makes the metal an essential nutrient for those life-forms. It is also used for plating and as a green tint in glass. In the laboratory, nickel is frequently used as a catalyst for hydrogenation. Nickel is often used in coins, or occasionally as a substitute for decorative silver.

Rechargeable nickel batteries are one type of alkaline storage cylindrical battery and classified as secondary batteries.

Nickel battery has a positive electrode made of active material-nickelous hydroxide. Because of the perfectly, sealed construction and the efficient charge/discharge characteristics, nickel batteries provide superior features and practical values in long service life, high-rate discharge and stable performance. As a result, they are widely used in many fields such as communication and telephone equipment, office equipment, tools, toys and emergency devices and consumer applications.

A typical jet engine today contains about 1.8 tonnes of nickel alloys and includes a long list of tailor-made nickel-based-alloys to meet specific needs (Nickel Magazine 2007). Pure nickel is a strong candidate for protective coating in bio-diesel storage applications, due to its high resistance to the corrosive nature of biodiesel and its vapors and minimal catalytic effects on the oxidation of biodiesel (Boonyongmaneerat et al. 2011).

1.4.8 Copper

Copper is a ductile metal with very high thermal and electrical conductivity. Pure copper is rather soft and malleable, and a freshly exposed surface has a pinkish or peachy color. It is used as a thermal conductor, an electrical conductor, a building material, and a constituent of various metal alloys.

Copper is the most widely used metal because of high conductivity. Copper and copper-based alloys are unique in their physical and mechanical properties. They have excellent corrosion resistance, high resistance to fatigue and relative ease of joining by soldering available in wide variety of forms (Pillai 2007).

Copper is easily worked, being both ductile and malleable. The ease with which it can be drawn into wires makes it useful for electrical work in addition to its excellent electrical properties. Copper can be machined, although it is usually necessary to use an alloy for intricate parts, such as threaded components, to get really good machinability characteristics. Good thermal conduction makes it useful for heat sinks and in heat exchangers. Copper has good corrosion resistance; it has excellent brazing and soldering properties and can also be welded, although best results are obtained with gas metal arc welding (Sambamurthy 2007). Copper as both metal and pigmented salt has a significant presence in decorative art.

1.4.9 Zinc

Zinc compounds are actively investigated because of their significant properties. Zinc oxide, being an n-type semiconductor with a wide direct gap of about 3.2 eV, has received much attention as a low-cost material for transparent and conductive films (Futsuhara et al. 1998). Zinc phosphide a II–V compound exists as a p-type semiconductor with a direct gap of near 1.51 eV, and is a promising low-cost material for solar cells due to its band structure (Pawlikowski 1981).

Zinc provides immunity, fertility and the capacity of senses including sight, taste and smell, notes the International Zinc Association. Zinc can also be recycled indefinitely, without losing any of its structural or functional characteristics (http://www.livestrong. com/article/199141-uses-for-zinc-powder/).

Zinc-air batteries (non-rechargeable) and zinc-air fuel cells (mechanically rechargeable) are electro-chemical batteries powered by oxidizing zinc with oxygen from the air. These batteries have high-energy densities and are relatively inexpensive to produce. Sizes range from very small button cells for hearing aids, larger batteries used in film cameras that previously used mercury batteries, to very large batteries used for electric vehicle propulsion.

In operation, a mass of zinc particles form a porous anode, which is saturated with an electrolyte. Oxygen from the air reacts at the cathode and forms hydroxyl ions which migrate into the zinc paste and form zincate, releasing electrons to travel to the cathode. The zincate decays into zinc oxide and water returns to the electrolyte. The water and hydroxyls from the anode are recycled at the cathode, so the water is not consumed. The reactions produce a theoretical 1.65 V, but this is reduced to 1.4–1.35 V in available cells. Zinc-air batteries have some properties of fuel cells as well as batteries, with zinc as the fuel, the reaction rate can be controlled by varying the air flow, and oxidised zinc/electrolyte paste can be replaced with fresh paste. Metallic zinc could be used as an alternative fuel for vehicles, in a zinc-air battery (Noring et al. 1993). Zinc-air batteries are considerably more safer in combating situations and more environmental friendly than lithium batteries (http://www.defense-update. com/products/z/zinc-air-battery-new.htm).

A Switzerland-based company, ReVolt uses zinc-air battery technology for hearing aids. ReVolt's battery claims to store three times more energy than lithium–ion by volume, and could incur just half the costs (http://www.goodcleantech. com/2009/11/a_powerful_rechargeable_zinc-a.php).

1.4.10   Tin

As the trend towards further miniaturisation of electronic products continues apace, packaging technology has progressed from the conventional wire and tape automated bonding to area array flip-chip bonding, which is able to provide increased input/output (I/O) counts and improved electrical performance (Qin et al. 2010). The advantages of this technology include high-density bonding, improved self-alignment, reliability and ease of manufacture (Wolf et al. 2006). One major step in the flip-chip interconnection process routes involves the deposition of, normally, solder alloys onto the bond pads of the chips (also known as solder bumping). With respect to the bumping materials, lead–tin-based alloys were the most widely used solders for flip-chip applications because of their low cost, low-melting point and excellent solderability properties. However, with world-wide legislation for the removal/reduction of lead and other hazardous materials from electrical and electronic products, development of a large number of lead-free, mostly tin-rich, alternative solders has been undertaken (Eveloy et al. 2005). Typically containing more than 90 wt% Sn, with a wide range of alloying elements such as Ag, Cu, In, Bi and Zn, these lead-free alternatives can be binary, ternary and even quaternary alloys, with variations in compositions. Sn–Ag–Cu solders can promote enhanced joint strength and creep and thermal fatigue resistance, and permit increased operating temperatures for advanced electronic systems and devices (Fabio and Mascaro 2006).

In electronic/optoelectronic packaging, chip bonding serves three major functions, i.e., mechanical support, heat dissipation and electrical connection (Hunziker et al. 1996). The choice of solder material for bonding is based on optimisation of a number of properties, including solderability, melting temperature, Young's modulus (or stiffness), coefficient of thermal expansion, Poisson's ratio, fatigue life, creep rate and corrosion resistance. In terms of melting temperature, solders are typically classified as either hard (high-melting temperature) or soft (low-melting temperature). The Pb/Sn system is an example of a soft solder, which is commonly used for electronic packaging. Hard solders, e.g., Au/Sn, are used for optoelectronic packaging. Au/Sn solder, with its combination of good thermal and electrical conductivities, is particularly attractive for 'flip-chip' bonding, where the active area of the device is next to the submount. Au-20 wt% Sn is the most common composition utilised; it has a relatively high-melting temperature (280°C), good creep behaviour and good corrosion resistance (Ivey 1998).

Guide wires, catheters, stents, etc., are being increasingly employed in the diagnosis and treatment of cancer, diseases of the circulatory system, etc. A guide wire is used for navigating a catheter, a tube made of plastic, in a blood vessel. The tip portion of the guide wire must be sufficiently flexible to pass through the meandering blood vessels. On the other hand, in the body portion of the guide wire, a high-elastic modulus and strength against bending are also required to overcome the high resistance to bending and rotation in a blood vessel and to smoothly transmit the torque from the end to the tip of the guide wire (Sutou et al. 2006). Ti–Mo–Sn alloy is found to be a promising biocompatible material for use in catheters (Maeshima and Nishida 2004).

1.4.11 Tellurium

Due to the remarkable physical properties of tellurium such as low band gap and transparency in the infrared region, Te is used extensively in various technological areas. Te thin films find use in microelectronic devices such as gas sensor (Shashwati et al. 2004; Tsiulyanu et al. 2004) and optical information storage (Josef et al. 2004). Tellurium-based thin films, suitable for applications in environmental monitoring with considerably short-response time and high sensitivity to nitrogen dioxide at room temperature have been reported (Tsiulyanu et al. 2001).

Tellurium, with a low band gap of 0.32 eV, is one of the most promising materials for a shield in a passive radiative cooling (Engelhard et al. 2000). Radiative cooling is the one among today's challenges in materials science research. It occurs when a body gets cold by loosing energy through radiative processes. The phenomenon of radiative cooling uses the fact that the thermal energy emitted by a clear sky in the "window region" (8–13 mm) is much less than the thermal energy emitted by a blackbody at ground air temperature in this wavelength range. Hence, a surface on the earth facing the sky experiences an imbalance of outgoing and incoming thermal radiation and cools to below the ambient air temperature. While this concept can work well at night, assuming a relatively dry atmosphere, the solar energy input during the day, which is normally much greater than that radiated out, causes heating of the system. To prevent this, a shield is required to cover the radiating surface in order to block solar radiation during the day as well as to prevent convective mixing in the cooled space. An ideal radiation shield should completely reflect solar radiation, but allow complete transmission in the "atmospheric-window" region. Solar radiation should preferably be reflected, as any absorbed radiation will be converted to heat somewhere in the system (Dobson et al. 2003). The different approaches for the design of shield are the introduction of optical scattering materials into shield substrates, and coating of shield substrates with high-solar reflector films. Te thin films with thickness of 111–133 nm having high-IR transmission across the full 8–13 μm band region suitable for solar radiation shield devices have been prepared by chemical vapor deposition method (Tian et al. 2006). Tellurium is used in photocopiers to enhance picture quality.

1.5 Significance of Alloys

Technology is reshaping our day-to-day life. Metals and their alloys make today's manufacturing industry, agriculture, construction and communication systems, transportation, defense equipments, etc., possible. Our manufacturing industries are using different metals and alloys as raw materials for their finished goods.

There are a large number of possible combinations of different metals and each has its own specific set of properties. Alloying one metal with other metal or non-metal often enhances its properties. For example, steel is stronger than iron, its primary element. The physical properties, such as density, reactivity, Young's modulus and electrical and thermal conductivity, of an alloy may not differ greatly from those of its elements, but engineering properties, such as tensile strength (Francis 2008) and shear strength may be substantially different from those of the constituent materials. This is sometimes due to the size of the atoms in the alloy, since larger atoms exert a compressive force on neighboring atoms, and smaller atoms exert a tensile force on their neighbors, helping the alloy resist deformation. Sometimes alloys may exhibit marked differences in Behaviour even when small amounts of one element occur (Hogan 1969; Zhang and Suhl 1985). Some of the major reasons for the continuing advances in alloys are the availability of materials, new manufacturing techniques, and the ability to test alloys before they are ever produced. Most modern alloys are, in fact, preplanned using sophisticated computer simulations, which help determine what properties the alloy will display.

1.5.1 Alloys in Nuclear Reactors

Structural materials employed in reactor systems must possess suitable nuclear and physical properties and must be compatible with the reactor coolant under the conditions of operation. The most common structural materials employed in reactor systems are stainless steel and zirconium alloys. Zirconium alloys have favourable nuclear and physical properties (Kutty et al. 1999), whereas stainless steel has favourable physical properties. Aluminium is widely used in low-temperature test and research reactors; zirconium and stainless steel are used in high-temperature power reactors. Zirconium is relatively expensive, and its use is therefore confined to applications in the reactor core where neutron absorption is important.

1.5.2 Alloy Wheels

Alloy wheels are automobile (car, motorcycle and truck) wheels which are made from an alloy of aluminium or magnesium (or sometimes a mixture of both). They are typically lighter for the same strength and provide better heat conduction and improved cosmetic appearance. Lighter wheels can improve handling by reducing vehicle mass which helps to reduce fuel consumption. Better heat conduction can help dissipate heat from the brakes, which improves braking performance in more demanding driving conditions and reduces the chance of brake failure due to overheating (Nunney 2006). Alloy wheels are also purchased for cosmetic purposes as the alloys used are largely corrosion-resistant. This permits the use of attractive bare-metal finishes, with no need for paint or wheel covers, and the manufacturing processes allow intricate, bold designs. Magnesium alloy wheels, or "mag wheels" are sometimes used on racing cars, in place of heavier steel or aluminium wheels, for better performance.

1.6 Significance of the Alloys Dealt With in this Research Work

Some of the commercially and industrially important metals and alloys have been analyzed in this work and their significance are discussed briefly in the following lines.

1.6.1 Cobalt Aluminium

Intermetallic alloys, including CoAl and NiAl, are of great importance since these materials not only have good strength-to-weight ratio but also has excellent corrosion and oxidation resistance, which make them good candidates for high-temperature and soft magnetic applications (Wan et al. 2010). Among the intermetallics, Cobalt aluminides are of a considerable technological interest for high-temperature applications (Lee et al. 2004). Cobalt forms a stable B2 (CsCl) structure with Al in a wide concentration range (Botton et al. 1996). This structure can be considered as two interpenetrating primitive cubic sublattices, where each Co atom has eight Al atoms as nearest neighbors and vice versa. Depending on the concentration, CoAl alloys exhibit different mechanical and magnetic properties (Kudryavtsev et al. 1998). CoAl is especially attractive for epitaxial growth due to its low lattice mismatch with respect to GaAs (Wan et al. 2010). In combination with semiconductor hetero structures, they are able to be integrated in a number of devices, e.g., spin injectors and mirrors.

Cobalt-based super alloys consume most of the produced cobalt. The temperature stability of these alloys makes them suitable for use in turbine blades for gas turbines and jet aircraft engines. Cobalt-based alloys are also corrosion and wear-resistant. Cobalt-based alloys are used in total joint replacement applications where high strength and corrosion resistance are necessary. Special cobalt–chromium–molybdenum alloys are used for prosthetic parts such as hip and knee replacements (Michel et al. 1991). Cobalt alloys are also used for dental prosthetics, where they are useful to avoid allergies to nickel. Some high-speed steels also use cobalt to increase heat and wear-resistance. The special alloys of aluminium, nickel, cobalt and iron, known as Alnico, are used in permanent magnets (Luborsky et al. 1957).

1.6.2 Nickel Aluminium

NiAl, an intermetallic compound, is a promising material for aerospace applications. It has a combination of important structural properties, such as high mechanical strength, low density, high-melting point, high thermal conductivity and excellent oxidation resistances (Choudry et al. 1998; Ng et al. 1997). It has potential applications such as hot sections of gas turbine engines for aircraft propulsion systems, coats under thermal barrier coating, electronic metallisation compounds in advanced semiconductors and surface catalysts (Albiter et al. 2002). Strong bonding between aluminium and nickel, which persists at elevated temperatures yields excellent high-temperature properties and specific strength that are competitive with those of super alloys and ceramics. Thus, these alloys offer new opportunities for applications in gas turbines at temperatures higher than those currently possible with conventional nickel-based super alloys. The advantages of NiAl-based intermetallics are a high-melting point above 1,460°C and high thermal conductivity. Due to their excellent high-temperature properties NiAl-based intermetallics are potential candidate materials for combustion chambers heat shields and first row vanes in industrial gas turbines (Scheppe et al. 2002). Polycrystalline NiAl exhibits a brittle–ductile transition at temperatures ranging from 300 to 600°C which are significantly lower than those of other intermetallic compounds. These properties have made NiAl-based alloy a promising candidate in some high-temperature (HT) structural applications (Gaoa et al. 2005).

Fine grains of a nickel-aluminium alloy, known as Raney nickel are used in many industrial processes. It is used as a heterogeneous catalyst in a variety of organic syntheses, most commonly for hydrogenation reactions. Its structural and thermal stability (i.e., the fact that it does not decompose at high temperatures) allows its use under a wide range of reaction conditions (Carruthers 1986). Raney nickel is used in a large number of industrial processes and in organic synthesis because of its stability and high-catalytic activity at room temperature (Hauptmann and Walter 1962).

1.6.3 Nickel Chromium

Nichrome wire, an alloy typically made of 80% nickel and 20% chrome, has specific characteristics that make it ideal for certain uses. It is silvery-grey in color, is corrosion resistant and has a high-melting point of about 1,400°C. Its ability to conduct electricity, withstand heat, its strength, flexibility and melting point determine its uses and applications.

Nichrome wire is used to make coils for heating elements. Due to its relatively high resistivity and resistance to oxidation at high temperatures, it is widely used in heating elements, such as in hair dryers, electric ovens, toasters and irons. Industrial uses include a wide variety of heating elements for ovens of all sizes, and devices that must apply heat directly to surfaces, such as for sealing plastic packages. Medical laboratories use nichrome wire loops somewhat like small spoons for handling specimens because they withstand frequent and repeated sterilisation, then cool rapidly. In industrial use, fine nichrome wire meshes serve as filters for liquids at high temperatures. Since the wire heats up rapidly with even a small amount of electricity, it makes safe igniters (sometimes called electric matches) to light the fuses of fireworks and model rockets. A piece of nichrome wire about an inch-and-a-half long connected to a length of lead wire allows the operator to safely detonate the device from a distance with only a battery and a switch. Fireworks professionals use nichrome igniters with timers at large displays. NiCr thin films are widely used in several applications in microelectronics such as thin film resistors, filaments and humidity sensors because of their relatively large resistivity, more resistant to oxidation and a low-temperature coefficient of resistance (Kwon et al. 2005). $Ni_{80}Cr_{20}$ matrix offers excellent high-temperature oxidation/corrosion resistance and essential mechanical strength (Ding et al. 2007). For dental clinical applications, the NiCr-based casting alloys are developed as an alternative to gold-based alloys due to their corrosion resistance in oral environment (Huang 2003). Nichrome $Ni_{80}Cr_{20}$ wt% films are used for strain gauge applications because of their high resistivity, low-temperature coefficient of resistance, commercial availability and low-temperature dependence of gauge factor (Chen et al. 2008).

1.6.4 Iron–Nickel

Iron group metals and binary alloys have a number of important industrial applications (Fabio and Mascaro 2006). Fe–Ni alloys have attracted much attention due to their interesting mechanical and magnetic properties (Karayannis and Moutsatsou 2006). Permalloy is a Fe–Ni alloy used in soft magnetic read/write heads (Cacciamani et al. 2010). Fe–Ni-based alloy powders are interesting in their applications as soft magnetic materials with low coercivity and high permeability (Gheisari et al. 2009).

One of the important challenges of electrical motors is to increase the efficiency to obtain more power with the same electrical energy consumed. The use of iron–nickel alloys increases the efficiency by reducing the magnetic losses significantly (Frederic et al. 2007). Fe–Ni thin films and multilayered structures are materials for high-frequency devices, such as inductors or magneto impedance effect (MI)-based magnetic field sensors (Kurlyandskaya et al. 2011).

Typical applications that are based on the low coefficient of thermal expansion of Fe–Ni alloys include thermostatic bimetals, glass sealing, integrated circuit packaging, cathode ray tube shadow masks, composite molds/tooling and membranes for liquid gas tankers (Cacciamani et al. 2010). Applications based on the soft magnetic properties include read-write heads for magnetic storage, magnetic actuators, magnetic shielding and high-performance transformer cores. Due to their unique low coefficient of thermal expansion and soft magnetic properties, Fe–Ni alloys are used in several industrial applications (McCrea et al. 2003).

1.6.5 Sodium Chloride Doped with Silver

Silver alkali halides provide interesting model systems for the study of decomposition processes in ionic solids. This is not only due to the absence of structural phase transitions but also due to the invariance of the anion sublattice, which is not involved in the demixing process. The phase separation is entirely confined to the cationic system and the anions exhibit an almost rigid frame. Along with the strong polarizability of silver ions, this feature guarantees that even single crystals are not destroyed during demixing (Elter et al. 2005). The dynamics of $Na_{1-x}Ag_xCl$ show that in the homogeneous phase, the doping of NaCl with silver ions leads to a considerable softening of the lattice (Caspary et al. 2007).

Semiconductors in confined surroundings experience a very significant development considering their importance in optoelectronic technologies. The semiconductors concerned are those having a direct and wide band gap conferring them with a radiative character and consequently a rather considerable output of photoluminescence, contrary to indirect gap semiconductors with a weaker output. Photosensitivity is observed in ionic semiconductors such as sodium chloride doped with silver (Madani et al. 2004).

1.6.6 Aluminium Doped with Iron

The addition of iron as a dopant in aluminium for integrated circuit applications substantially increases resistance to electro migration and creep. The amount of iron utilised depends to an extent on the electrical requirements of the device, the geometry of the device, the substrate composition and composition of overlying layers.

Stress-induced grain boundary movement in aluminium lines used as connections in integrated circuits are substantially avoided by doping aluminium with iron. Through this expedient not only is grain boundary movement is avoided but the electro migration problems are decreased (Ryan et al. 1993).

Aluminium doped with iron is distinguished by high thermal-shock resistance and, at temperatures of $800\,^{\circ}C$ has comparatively good mechanical properties. It has mechanical properties which permit its use in components which are slightly stressed mechanically. It has excellent shock resistance and can therefore be used in those parts of thermal installations which are subject to frequent thermal cycling, such as in particular as a casing or casing part of gas turbine or of turbo charger or as a nozzle ring (Nazmy et al. 1995).

To meet the recent trends towards appealing, custom effects, many new effect pigments have been developed in the last few years in the automotive industry in recent past. Besides micas, flake-like particles have been doped with ultra thin layers of metal oxides and launched into coating markets. Aluminium flakes, doped with thin layers of iron oxide find much demand in this industry (Poth 2008).

1.7 Ball Milling

Most crystalline solids are composed of a collection of many small crystals or grains, termed polycrystalline. The term nano crystalline materials (Gleiter 1989) are used to describe those materials that have a majority of grain diameters in the typical range from 1 to 50 nm (McHenry and Laughlin 2000). The nano crystalline materials have received much attention as advanced engineering materials with unique physical and mechanical properties. The mechanical properties of the nano crystalline materials at room temperature have higher strength and toughness to those of coarse-grained ones (Gleiter 1981).

Nano crystalline materials can be successfully synthesised by several techniques, including inert gas condensation (Birringer et al. 1984), rapid solidification (Inoue 1994), sputtering (Li and Smith 1989), crystallisation of amorphous phases (Lu et al. 1995) and chemical processing (Kear and Strutt 1995). Among the different options for preparations, the ball milling method has been considered the most powerful tool for nanostructure materials because of its simplicity relatively inexpensive equipment and the possibility of producing large quantities that can be scaled up to several tons (Zhao et al. 2010).

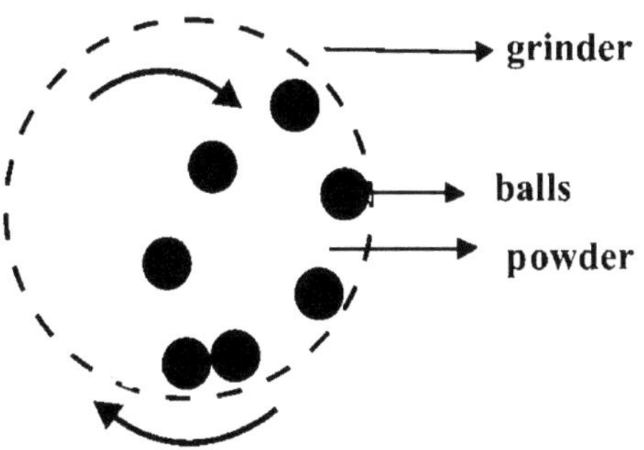

Fig. 1.2 Balls and powder sample in a ball mill

A ball mill is a type of grinder and a cylindrical device used in grinding (or mixing) materials like ores, chemicals and ceramic raw materials. Ball mills rotate around a horizontal axis, partially filled with the material to be ground and the grinding medium (Fig. 1.2). Different materials are used as media, including ceramic balls, alumina balls and stainless steel balls. An internal cascading effect reduces the material to a fine powder. Industrial ball mills can operate continuously fed at one end and discharged at the other end. Large-to-medium sized ball mills are mechanically rotated on their axis, but small ones normally consist of a cylindrical capped container that sits on two drive shafts (pulleys and belts are used to transmit rotary motion) (Rajagopal 2009).

Ball mills are also used in pyrotechnics and the manufacture of black powder, but cannot be used in the preparation of some pyrotechnic mixtures such as flash powder because of their sensitivity to impact. High-quality ball mills are potentially expensive and can grind mixture particles to as small as 5 nm, enormously increasing surface area and reaction rates. The grinding works on principle of critical speed. The critical speed can be understood as the speed after which the balls (which are responsible for the grinding of particles) start rotating along the direction of the cylindrical device; thus causing no further grinding (Malik and Singh 2010). Ball mills are used extensively in the Mechanical alloying process in which they are not only used for grinding but also for cold welding as well, with the purpose of producing alloys from powders.

1.7.1 Mechanism for the Formation of Nano Crystalline Materials by the Ball Milling

The mechanism for formation of nano crystalline materials by the ball milling technique has been summarised as the phenomenology of the grain size reduction into three stages (Fecht 1995).

First stage: Plastic deformation is produced in the crystal lattices of the ball-milled powders by slip and twinning. This deformation is localised in shear bands containing a high-dense network of dislocation. The local shear instability of a crystal lattice can be triggered by the material heterogeneity and enhance instabilities. These instabilities result from a non-uniform heat transfer during the mechanically induced deformation of the milled powders. During this stage of milling, the atomic level strain increases as a result of increasing the dislocation density.

Second stage: Due to the successive accumulation of the dislocation density, the crystals are disintegrated into sub-grains that are initially separated by low-angle grain boundaries. The formation of these sub-grains is attributed to the decrease of the atomic level strain.

Third stage: Further ball milling time leads to further deformation occurring in the shear bands located in the unstrained parts of the powders which leads to sub-grain size reduction so that the orientation of final grains become random in crystallographic orientations of the numerous grains and hence, the direction of slip varies from one grain to another.

In ball mills, the useful kinetic energy can be applied to the powder particles of the reactant materials by

- Collision between the balls and the powders (Fig. 1.2)
- Pressure loading of powders pinned between milling media or between the milling media and the liner
- Impact of the falling milling media
- Shear and abrasion caused by dragging of particles between moving milling media
- Shock wave transmitted through crop load by falling milling media.

1.7.2 Effect of Materials of Milling Media

There are many types of milling media suitable for use in a ball mill, each material having its own specific properties and advantages (Rajagopal 2009). Common in some applications are stainless steel balls. While usually very effective due to their high density and low contamination of the material being processed, stainless steel balls are unsuitable for some applications, such as black powder and other flammable materials require non-sparking lead, antimony, brass or bronze grinding media. In some applications ceramic or flint grinding media is used. Ceramic media are also very resistant to corrosive materials. High-density alumina media are widely used to grind clay bodies, frits, glazes and other ingredients. It is more expensive than silica media but is more efficient.

Fig. 1.3 Laboratory ball mill

1.7.3 Laboratory Ball Mill

The laboratory ball mill is a low-energy ball mill which is used in the present study. It is less expensive and operates with minimum maintenance requirements. The vial is a cylinder made of stainless steel with a radius of 2.5 cm. Stainless steel balls of different radii have been used as the milling media. The rotation speed is about 200 rpm. It produces homogenous and uniform powders. The laboratory ball mill used in our work is shown in Fig. 1.3.

1.8 X-Ray Diffraction

The modern understanding of metals and alloys, their structures, defects and various properties would not be possible if their crystal structures had not been revealed by XRD studies. XRD is a non-destructive technique for analyzing a wide range of materials, including crystals, metals, minerals, polymers, thin film coating and ceramics. Crystalline materials are characterised by the orderly, periodic arrangements of atoms. The unit cell is the basic repeating unit that defines a crystal. Parallel planes of atoms intersecting the unit cell are used to define directions and distances in the crystal. These crystallographic planes are identified by miller indices (Warren 1990). Diffraction from different planes of atoms produces a diffraction pattern, which contains information about the atomic arrangement within the crystal.

XRD is based on constructive interference of monochromatic X-rays and a crystalline sample. These X-rays are generated by a cathode ray tube, filtered to produce monochromatic radiation, collimated to concentrate and directed towards the sample. The interaction of the incident rays with the sample produces constructive interference (and a diffracted ray) when conditions satisfy Bragg's Law ($n\lambda = 2d \sin \theta$) (Bragg 1912). This law relates the wavelength of

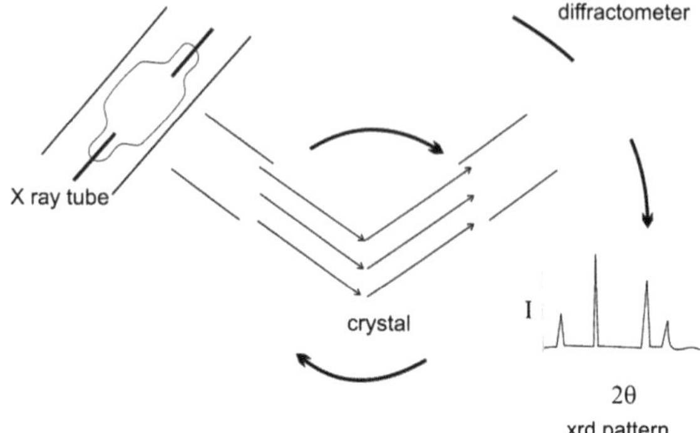

Fig. 1.4 Determination of crystal structure through X-rays

electromagnetic radiation to the diffraction angle and the lattice spacing in a crystalline sample. These diffracted X-rays are then detected, processed and counted.

Figure 1.4 illustrates how crystalline structure may be determined through XRD. As the crystal and detector rotate, X-rays diffract at specific angles. The detector reports the intensity (I) of X-ray photons as it moves. Angles of diffraction (where the Bragg equation is satisfied) are marked by peaks. The peak height is a function of the interaction of the X-rays with the crystal and the intensity of the source.

1.8.1 X-Ray Diffraction Methods

Classically, the two main ways of studying metals and alloys were metallography (the examination of polished and etched surfaces) and cooling curves (looking for discontinuities that indicated some sort of phase change). Both these methods involved considerable skill and experience, and the results were not always unambiguous. The introduction of XRD provided a much clearer, simpler and more objective way of investigation. XRD is now a common technique for the study of crystal structures and atomic spacing.

In terms of the specimen handled, two methods can be identified,

1. Powder diffraction
2. Single-crystal diffraction

In the former, the specimen is a collection of crystallites. Since these fragments are completely randomly oriented the incident X-ray beam meets with every possible lattice plane, oriented in all directions. Whereas in the single-crystal method, the whole specimen is a single piece, without any discontinuity in the lattice arrangements (Tareen and Kutty 2001). All diffraction methods are based on generation of X-rays in an X-ray tube. These X-rays are directed to the sample, and the diffracted rays are collected. A key component of all diffraction is the angle between the incident and diffracted rays.

1.8.2 Diffractometers

A typical X-ray diffractometer consists of a source of radiation (X-ray tube), a monochromator to choose the wavelength, slits to adjust the shape of the beam, sample and a detector. In a more complicated apparatus also a goniometer can be used for fine adjustment of the sample and the detector positions (Azaroff 1968).

1.8.3 Powder X-Ray Diffraction Instrumentation

The powder method essentially has two different ways of registering the diffracted X-rays (Tareen and Kutty 2001).

1. The whole diffraction pattern is recorded simultaneously on a photographic film called the powder photographic method.
2. The diffraction pattern is scanned by a counter device, or a solid-stated semi-conductor detector. The counter or detector registers the diffracted beam in successive stages, away from the direct beam.

In X-ray diffractometers, X-rays are generated in a cathode ray tube by heating a filament to produce electrons, accelerating the electrons towards a target by applying a voltage, and bombarding the target material with electrons. When electrons have sufficient energy to dislodge inner shell electrons of the target material, characteristic X-ray spectra are produced. These spectra consist of several components, the most common being K_α and K_β. K_α consists, in part, of $K_{\alpha 1}$ and $K_{\alpha 2}$. $K_{\alpha 1}$ has a slightly shorter wavelength and twice the intensity as $K_{\alpha 2}$ (Stout and Jensen 1989). The specific wavelengths are characteristic of the target material (Cu, Fe, Mo, and Cr). The monochromatic radiation required for the powder method is usually the $K_{\alpha 1 \alpha 2}$ doublet, monochromated by crystal reflection or by the use of a filter whose K absorption wavelength falls between the K_α and the K_β wavelengths. The Cu $K_{\alpha 1 \alpha 2}$ doublet ($\lambda = 1.542$ Å) with a Ni filter $\lambda_k = 1.488$ Å is probably used more than any other source (Warren 1990). These X-rays are collimated and directed onto the sample. The sample is mounted on a goniometer (Shirane et al. 2002) and gradually rotated while being bombarded with X-rays, producing a diffraction pattern of regularly spaced spots known as reflections. As the sample and detector are rotated, the intensity of the reflected X-rays is recorded. When the geometry of the incident X-rays impinging the sample satisfies the Bragg Equation, constructive interference occurs and a peak in intensity is seen. A detector records and processes this X-ray signal and converts the signal to a count rate which is then output to a device such as a printer or computer monitor.

1.8.4 Single-Crystal X-Ray Diffraction Instrumentation

The foremost essential criteria for single-crystal XRD are to obtain an adequate crystal of the material under study. The crystal should be sufficiently large (typically larger than 0.1 mm in all dimensions), pure in composition and regular in structure, with no significant internal imperfections such as cracks or twinning.

The crystal is placed in an intense beam of X-rays, usually of a single wavelength (monochromatic X-rays), producing the regular pattern of reflections. Molybdenum is the most common target material for single-crystal diffraction, with MoK_α radiation $= 0.7107$Å. These X-rays are collimated and directed onto the sample. When the geometry of the incident X-rays impinging the sample satisfies the Bragg Equation, constructive interference occurs. A detector records and processes this X-ray signal and converts the signal to a count rate which is then output to a device such as a printer or computer monitor. Modern single-crystal diffractometers use CCD (charge-coupled device) technology to transform the X-ray photons into an electrical signal which are then sent to a computer for processing.

Single-crystal diffractometers use either three- or four-circle goniometers. These circles refer to the four angles (2θ, χ, φ, and Ω) (Shirane et al. 2002) that define the relationship between the crystal lattice, the incident ray and the detector as shown in Fig. 1.5. Samples are mounted on thin glass fibers using an epoxy or cement. The thin glass fiber is attached to brass pins and mounted onto goniometer heads. The goniometer head and sample are then affixed to the diffractometer. Samples are centered by viewing the sample under an attached microscope and are placed under the cross-hairs for all crystal orientations. Once the crystal is centred, a preliminary rotational image is often collected to screen the sample quality and to select parameters for later steps. As the crystal is gradually rotated, previous reflections disappear and new ones appear, the intensity of every spot is recorded at every orientation of the crystal. Multiple data sets may have to be collected, with each set covering slightly more than half a full rotation of the crystal and typically containing tens of thousands of reflections (Massa 2004).

An automatic collection routine can then be used to collect a preliminary set of frames for determination of the unit cell. Reflections from these frames are auto-indexed to select the reduced primitive cell and calculate the orientation matrix (which relates the unit cell to the actual crystal position within the beam). These data are combined computationally with complementary chemical information to produce and refine a model of the arrangement of atoms within the crystal and converted to the appropriate crystal system and Bravais lattice (Giacovazzo 2002). The final, refined model of the atomic arrangement is the essential crystal structure.

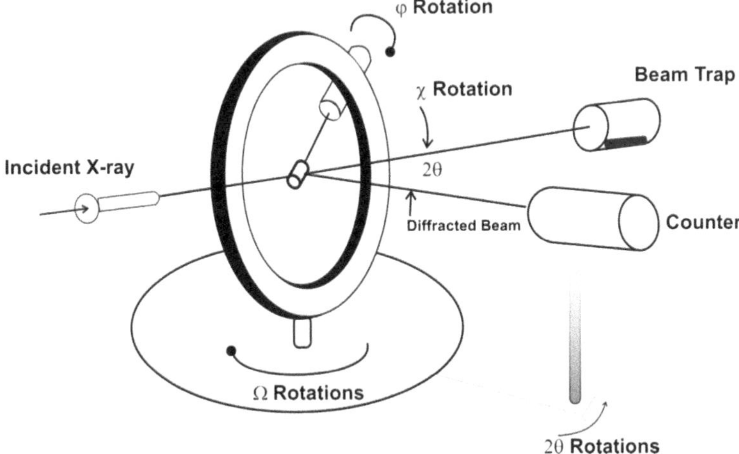

Fig. 1.5 Goniometer

1.8.4.1 Corrections for Background, Absorption

After the data have been collected, corrections for instrumental factors, polarisation effects of X-ray absorption (Glusker et al. 1994) and (potentially) crystal decomposition must be applied to the entire data set. This integration process also reduces the raw frame data to a smaller set of individual integrated intensities. These correction and processing procedures are typically part of the software package which controls and runs the data collection.

Single-crystal XRD is most commonly used for precise determination of a unit cell, including cell dimensions and positions of atoms within the lattice. Bond-lengths and angles are directly related to the atomic positions. A single, robust, optically clear sample, generally between 50 and 250 microns in size is essential. Data collection generally requires between 24 and 72 h.

1.9 Grain Size Analysis from X-Ray Diffraction

The size of the grains in a polycrystalline material has more effect on the properties of the material, for example, the hardness of a metal or alloy increases with decrease in the grain size (Cullity and Stock 2001). This dependence of properties on grain size makes the measurement of grain size important in the control of most metal forming operations.

In this book the grain size has been reported for some metals which has been analyzed from XRD. The grain morphology was examined by SEM. The software GRAIN written by Dr.R.Saravanan was used to estimate approximate grain sizes from XRD (Saravanan, GRAIN software). The grain size is analyzed using full width at half maximum of the powder XRD peaks. The Debye–Scherrer formula given in Eq. 1.1has been used to calculate the particle size.

$$\tau = \frac{K\lambda}{\beta \cos \theta} \tag{1.1}$$

where K is the shape factor, λ is the X-ray wavelength, typically 1.54 Å, β is the line broadening at half the maximum intensity (FWHM) in radians and θ is the Bragg angle (Patterson 1939) τ is the mean size of the ordered (crystalline) domains, which may be smaller or equal to the grain size. The dimensionless shape factor has a typical value of about 0.9, but varies with the actual shape of the crystallite. The Scherrer equation is limited to nano-scale particles. It is not applicable to grains larger than about 0.1 µm, which precludes those observed in most metallographic and ceramographic microstructures.

It is important to realise that the Scherrer formula provides a lower bound on the particle size. The reason for this is that a variety of factors can contribute to the width of a diffraction peak; besides particle size, the most important of these are usually inhomogeneous strain and instrumental effects. If all of these other contributions to the peak width were zero, then the peak width would be determined solely by the particle size and the Scherrer formula would apply. If the other contributions to the width are non-zero, then the particle size can be larger than that predicted by the Scherrer formula, with the "extra" peak width coming from the other factors.

1.10 Scanning Electron Microscope

The SEM uses a focused beam of high-energy electrons to generate a variety of signals at the surface of solid specimens (Malik and Singh 2010). The signals derived from electron-sample interactions reveal information about the sample including external morphology (texture), chemical composition and crystalline structure and orientation of materials making up the sample. In most applications, data are collected over a selected area of the surface of the sample, and a two-dimensional image is generated that displays spatial variations in these properties. Areas ranging from approximately 1 cm to 5 microns in width can be imaged in a scanning mode using conventional SEM techniques (magnification ranging from 20X to approximately 30,000X, spatial resolution of 50–100 nm). The SEM is also capable of performing analyses of selected point locations on the sample; this approach is especially useful in qualitatively or semi-quantitatively determining chemical compositions, crystalline structure and crystal orientations (Reimer 1998).

SEM is also widely used to identify phases based on qualitative chemical analysis and/or crystalline structure. Backscattered electron images (BSE) can be used for rapid discrimination of phases in multiphase samples. SEMs equipped with diffracted backscattered electron detectors (EBSD) can be used to examine microfabric and crystallographic orientation in many materials.

1.11 Fundamental Principles of Scanning Electron Microscopy

Accelerated electrons in SEM carry significant amounts of kinetic energy, and this energy is dissipated as a variety of signals are produced by electron-sample interactions when the incident electrons are decelerated in the solid sample. These signals include secondary electrons (that produce SEM images), backscattered electrons (BSE), diffracted backscattered electrons (EBSD that are used to determine crystal structures and orientations of minerals), photons (characteristic X-rays that are used for elemental analysis and continuum X-rays), visible light (cathodoluminescence-CL) and heat. Secondary electrons and backscattered electrons are commonly used for imaging samples, secondary electrons are most valuable for showing morphology and topography of samples and backscattered electrons are most valuable for illustrating contrasts in composition in multiphase samples (i.e., for rapid phase discrimination). X-ray generation is produced by inelastic collisions of the incident electrons with electrons in discrete orbitals (shells) of atoms in the sample. As the excited electrons return to lower energy states, they yield X-rays that are of a fixed wavelength (that is related to the difference in energy levels of electrons in different shells for a given element). Thus, characteristic X-rays are produced for each element in a mineral that is "excited" by the electron beam (Goldstein 2003).

SEM analysis is considered to be "non-destructive", i.e., X-rays generated by electron interactions do not lead to volume loss of the sample, so it is possible to analyze the same materials repeatedly.

References

Albiter A, Bedolla E, Perez R (2002) Mater Sci Eng A 328:80–86
Almanza R, Hernández P, Martínez I, Mazari M (2009) Sol Energ Mater Sol Cells 93:1647–1651
Ares JR, Guégan P, Cuevas FA (2005) Acta Materialia 53.2157–2167
Audi G (2003) Nucl Phys A (Atomic Mass Data Center) 729:3–128
Auxer W (1986) Electrochemical Society, pp 49–54
Azaroff LV (1968) Elements of X-ray crystallography, Chap. 13. McGraw Hill, New York
Birringer R, Gleiter H, Klein HP, Marquardt P (1984) Phys Lett A 102:356–369
Boonyongmaneerat Y, Sukjamsri C, Sahapatsombut U, Saenapitak S, Sukkasi S (2011) Appl Energ 88:909–913
Botton GA et al (1996) Phys Rev B 54:1682–1691
Bragg WL (1912) Nature 90:410–410
Cacciamani G, Dinsdale A, Palumbo M, Pasturel A (2010) Intermetallics 18:1148–1162
Carruthers W (ed) (1986) Some modern methods of organic synthesis. Cambridge University Press, Cambridge, pp 413–414

Caspary D, Eckold G, Elter P, Gibhardt H, Güthoff F, Demmel F, Hoser A, Schmidt W, Ding CH, Yang ZM, Zhang HT, Guo YF, Zhoua JN (2007) Compost A 38:48–352

Chen R, Zuo D, Sun Y, Li D, Lu W, Wang M (eds) (2008) Key engineering materials, vol 375. Trans Tech Publications, Switzerland, pp 690–694

Choudry MS, Dollar M, Eastman JA (1998) Mater Sci Eng A256:25–33

Collins DM (1982) Nature 298:49–51

Crans DC (2000) J Inorg Biochem 80:123–131

Crans DC, Smee JJ, Gaidamauskas E, Yang L (2004) Chem Rev 104:849–902

Cullity BD, Stock SR (2001) Elements of X-ray diffraction, Chap. 14, 3rd edn. Prentice Hall, Upper Saddle

Ding CH, Yang ZM, Zhang HT, Guo YF, Zhoua JN (2007) Compos Part A 38:348–352

Dobrzánski LA, Tánski T, Čížek L, Brytan Z (2007) J Mater Process Technol 192:567–574

Dobson KD, Hodes G, Mastai Y (2003) Sol Energ Mater Sol Cells 80:283–296

Dorward RC, Pritchett T (1988) Mater Des 9(2):63–69

El-Eskandarany MS (2001) Mechanical alloying for fabrication of advanced engineering materials, Chap. 1. William Andrew, New York

Elter P, Eckold G, Gibhardt H, Schmidt W, Hoser A (2005) J Phys Condens Matter 17:6559–6573

Engelhard T, Jones ED, Viney I, Mastai Y, Hodes G (2000) Thin Solid Films 370:101–105

Eveloy V, Ganesan S, Fukuda Y, Wu J, Pecht MG (2005) IEEE Trans Compon Packag Technol 28:884–894

Fabio RB, Mascaro LM (2006) Surf Coat Technol 201:1752–1756

Fecht HJ (1995) NanoStruc Mater 6:33–42

Francis J (ed) (2008) Philosophy of mathematics. Global publishing house, New York, p 56

Frederic B, Thierry W, Herve F (2007) Electrical insulation conference and electrical manufacturing expo, pp 394–401

Futsuhara M, Yoshioka K, Takai O (1998) Thin Solid Films 322:274–281

Gallagher R, Ingram P (eds) (2001) New coordination science, chemistry for higher tier ingram, 3rd edn. Oxford University Press, New York, p 56

Gaoa Q, Guob JT, Huaib KW, Zhanga JS (2005) Mater Lett 59:2859–2862

Gheisari KH, Javadpoura S, Oh JT, Ghaffari M (2009) J Alloys Compd 472:416–420

Giacovazzo C (2002) Fundamentals of crystallography issue 7 of international union of crystallography texts Oxford science publication, Chap. 5, 2nd edn. Oxford University Press, Oxford

Gleiter H (1981) Deformation of polycrystals. In: Hansen N, Horsewell A, Leffers T, Lilholt H (eds) Proceedings second riso intrnational symposium metallurgy and materials science, Roskilde, p 15

Gleiter H (1989) Prog Mat Sci 33:223–315

Glusker JP, Lewis M, Rossi M (eds) (1994) Crystal structure analysis for chemists and biologists, volume 16 of methods in stereochemical analysis. Wiley, New York, p 261

Goldstein J (2003) Scanning electron microscopy and X-ray microanalysis, Chap. 6. Kluwer Adacemic/Plenum Pulbishers, New York

Hansen NK, Coppens P (1978) Acta Cryst A34:909–921

Hauptmann H, Walter W (1962) Chem Rev 62:347–404

Heinrich G, Sgler T, Rosiwal SM, Singer RF, Ley L (1996) Diam Relat Mater 5:304–307

Hogan CM (1969) Phys Rev 188:870–874

Hort N, Huang Y, Fechner D, Störmer M, Blawert C, Witte F, Vogt C, Drücker H, Willumeit R, Kainer KU, Feyerabend F (2010) Acta Biomater 6:1714–1725

http://itp.nyu.edu/physcomp/uploads/battery_nimh.jpg)

http://en.wikipedia.org/wiki/File:Vanadinite2_sur_goethite_ (Maroc).jpg

http://image.made-in-china.com/2f0j00DvhtLbGPOekg/Grinding-Ceramic-Balls.jpg

http://www.shubhmets.in/full-images/663342.jpg

Huang HH (2003) Biomaterials 24:1575–1582

Hunziker W, Voyt W, Melchior H (1996) Proceedings of the 1996 IEEE 46th electronic components and technology conference, Orlando, Florida, pp 8–12

Inoue A (1994) Mater Sci Eng A 179:57–61

Ivey DG (1998) Micron 29(4):281–287

Josef P, Dana P, Jaroslav B et al (2004) Anal Appl Pyrolysis 71:739–746

Kainer KU, Sb (2000) International congress magnesium alloys and their application, Mnichov, pp 534–608

Karayannis VG, Moutsatsou AK (2006) J Mater Process Technol 171:295–300

Kartal G, Timur S, Urgen M, Erdemir A (2010) Surf Coat Technol 204:3935–3939

Kazacos MS, Menictas C (1997) Telecommunications energy conference, 19th international, pp 463–471

Kear BH, Strutt PR (1995) Nanostruct Mater 6:227–236

Kittel C (1996) Introduction to solid state physics, 7th edn. Wiley, India

Kittel C (2007) Introduction to solid state physics, Chap. 6, 7th edn. Wiley, India

Kudryavtsev YV et al (1998) J Appl Phys 83:1575–1581

Kurlyandskaya GV, Bhagat SM, Svalov AV, Fernandez E, Arribas AG, Barandiaran JM (2011) Solid State Phenom 168:257–260

Kutty TRG, Ravi K, Ganguly C (1999) J Nucl Mater 265:91–99

Kwon Y, Kim N, Choi G, Lee W, Seo Y, Park J (2005) Microelectron Eng 82:314–320

Lee YS, Lee CH, Kim KW, Shin HJ, Lee YP (2004) J Magn Magn Mater 272:2151–2153

Li ZG, Smith DJ (1989) Appl Phys Lett 55:919–921

Li SJ, Cui TC, Hao YL, Yang R (2008) A Biomater 4:305–317

Lide DR (ed) (1999) In: CRC handbook of chemistry and physics, 80th edn. CRC Press, New York, pp 4–32

Lonyuk B, Apachitei I, Duszczyk J (2007) Surf Coat Technol 201:8688–8694

Lu K, Wei WD, Wang JT (1995) Scr Metall Mater 24:2319–2323

Luborsky FE, Mendelsohn LI, Paine TO (1957) J Appl Phys 28:344–351

Lynch CT (1974) CRC handbook of materials science. Vol 1,CRC press

Madani S, Hocine K, Madjid B, Abdelhamid C, Mohamed AB, Sofiane M (2004) Physica E 23:217–220

Maeshima T, Nishida M (2004) Mater Trans 45:1096–1100

Malik HK, Singh AK (2010) Engineering physics, Tata McGraw-Hill, New Delhi, p 6

Markovsky PE, Semiatin SL (2010) J Mater Process Technol 210:518–528

Massa W (2004) Crystal structure determination, Chap. 6, 2nd edn. Springer, Berlin

McCrea JL, Palumbo G, Hibbard GD, Erb U (2003) Rev Adv Mater Sci 5:252–258

McHenry ME, Laughlin DE (2000) Acta Mater 48:223–238

Mehta DS, Masood SH, Song WQ (2004) J Mater Process Technol 156:1526–1531

Michel R, Nolte M, Reich M, Löer F (1991) Systemic effects of implanted prostheses made of cobalt–chromium alloys, Archives of orthopaedic and trauma surgery, 110:61–74

Monfared HH, Alavi S, Bikas R, Mahedpour M, Mayer P (2010) Polyhedron 29:3355–3362

Mordike BL, Ebert T (2001a) J Mater Process Technol 117:381–385

Mordike BL, Ebert T (2001) Mater Sci Eng A 302:37–45

Nazmy M, Noseda C, Staubli M (1995) Switzerland, United States (19), Patent Number 5411702, May 2

Newbury DE (1986) Advanced scanning electron microscopy and X-ray microanalysis, Chap. 2. Springer, Berlin

Ng C, Simkin BA, Crimp MA (1997) Mater Sci Eng A239:150–156

Nickel Magazine (2007), 22(2) March issue

Nithya D, Kalaiselvan R (2011) Physica B 406:24–29

Noring J et al (1993) In: Proceedings of the symposium on batteries and fuel cells for stationary and electric vehicle applications vols 93–98 of proceedings (electrochemical society), The Electrochemical Society, pp 235–236

Nowosielski R, Babilas R, Ochin P, Stoktosa Z (2008) Arch Mater Sci Eng 30:13–16

Nunney MJ (ed) (2006) Light and heavy vehicle technology, 4th edn. Butterworth-Heinmann, London, p 454

Oshima T, Kajitani M, Okuno A (2004) Int J Appl Ceram Technol 1269–1276

Paik J, Veen SVD, Duran A, Collette M (2005) Thin-Walled Str 43:1550

Patterson A (1939) Phys Rev 56:978–982

Pawlikowski M (1981) Infrared Phys 21:181–187

Petersson IU, Löberg JEL, Fredriksson AS, Ahlberg EK (2009) Biomaterials 30:4471–4479

Pillai SO (2007) Solid state physics, New age international (p) limited, New Delhi, Reprint, p 288

Poth U (2008) Automotive coatings formulation: chemistry, physics and practices European coatings tech files, Vincentz, p 152

Proffen TH, Billinge SJL (1999) J Appl Cryst 32:572–575

Qin Y, Wilcox GD, Liu C (2010) Electrochim Acta 56:183–192

Rajagopal K (2009) Textbook of Engineering Physics, Part II, Publisher-PHI learning Pvt. Ltd, New Delhi, p 168

Reimer L (1998) Scanning electron microscopy, physics of image formation and micro analysis, Chap. 1. Springer, Berlin

Rietveld HM (1969) J Appl Crystallogr 2:65–71

Ryan VW, Schutz RJ, Warren (1993) United States Patent (19), Patent No: 5243221 September 7

Sambamurthy K (2007) Pharmaceutical engineering, new age international, p 427

Saravanan R http:\\www.saraxraygroup.net. Grain software

Scheppe F, Sahm PR, Hermann W, Paul U, Preuhs J (2002) Mater Sci Eng A329:596–601

Schwingel D, Seeliger HW, Vecchionacci C, Alwes D, Dittrich J (2007) Acta Astronaut 61:326–330

Setoyama D, Matsunaga J, Muta H, Uno M, Yamanaka S (2004) J Alloys Compds 381:215–220

Shashwati S, Muthe KP, Joshi N (2004) Sens Actuators B 98:154–159

Shirane G, Shapiro SP, Tranquada JM (2002) Neutron scattering with a triple-axis spectrometer: basic techniques, Cambridge University Press, Cambridge, p 88

Skyllas-Kazacos M (2003) J Power Sour 124:299–302

Sparrow G (1999) Iron—the elements-group 1, the elements series. Marshall Cavendish, p 4

Stout GH, Jensen LH (1989) X-ray structure determination, Chap. 1, 2nd edn. Wiley-Interscience, New York

Suryanarayana C (2004) Mechanical alloying and milling, Chap. 4, vol 180 of Mechanical engineering. Marcel Dekker, New York

Sutou Y, Yamauchi K, Takagi T, Maeshima T, Nishida M (2006) Mater Sci Eng A 438:64–69

Tareen JAK, Kutty TRN (2001) A basic course in crystallography, chemistry series, Universities Press, p 149

Tian MY, Qing GZ, Hong XW, Jian H (2006) Trans Nonferrous Met Soc China 16:693–699

Tsiulyanu D, Mariana S, Miron V, Liess HD (2001) Sens Actuators B 73:35–39

Tsiulyanu D, Marian S, Liess HD, Eisele I (2004) Sens Actuators B 100:380–386

Walawalkar R, Apt J, Mancini R (2007) Energ Policy 35(4):2558–2568

Wan Q, Hey R, Trampert A (2010) 16th International conference on microscopy of semiconducting materials, IOP Publishing J Phys Conference Series 209, 012023

Wang RM, Eliezer A, Gutman E (2002) Mater Sci Eng A344:279–287

Warren BE (1990) X-ray diffraction, 2nd edn. reprint, Courier Dover Publications, New York, p 16

Wolf MJ, Engelmann G, Dietrich L, Reichl H (2006) Nucl Instrum Method Phys Res A 565:290–295

Zhang X, Suhl H (1985) Phys Rev A 32:2530–2533

Zhang Y, Liu X, Xie G, Yu L, Yi S, Hua M, Huang C (2010) Mater Sci Eng B 175:164–171

Zhao KY, Li CJ, Tao JM, Ng DHL, Zhu XK (2010) J Alloys Compds 504S:306–310

Zumdahl SS (2007) Chemical principles, 6th edn. Cengage Publications, p 33

Chapter 2
Charge Density Analysis from X-Ray Diffraction

Abstract This chapter provides a survey of current applications of X-ray diffraction techniques in crystal structure analysis, with some focus on recent advances that have been made in the scope and potential for carrying out crystal structure determination directly from diffraction data. The basic concepts of crystal structure analysis, Rietveld refinement, and the concepts used for the estimation and analysis of charge density in a crystal are discussed. The more reliable models for charge density estimation like multipole formalism and maximum entropy method are discussed in detail. The local structural analysis technique atomic pair distribution function is also discussed.

2.1 Introduction

Knowledge of the atomic scale geometrical structure of matter is a prerequisite for understanding and predicting the properties of technologically and scientifically important materials. The geometrical structure of a material not only consists of the time and space averaged periodic conformation of atoms in an idealized crystal lattice but also the microstructure caused by imperfections, dislocations, and all kinds of disorders that are often responsible for interesting properties of the material under investigation (Dinnibier 2008).

Establishing the structural properties of metals and alloys is essential for understanding their function and properties. In order to understand the properties of materials and to improve them, the atomic structure has to be known. The advent of X-ray methods to analyze metals was a great boon to metallurgists (Weiss 1925). The introduction of X-ray diffraction provided a much clearer, simpler, and more objective way of investigation.

Modern crystallography is not only about knowing the positions of the atoms and the corresponding bond distances, angles, and related features within a material, but the most important and interesting aspect of this technique is the

R. Saravanan and M. Prema Rani, *Metal and Alloy Bonding: An Experimental Analysis*,   31
DOI: 10.1007/978-1-4471-2204-3_2, © Springer-Verlag London Limited 2012

correlation of structural features with physical properties. This can include optical, magnetic, and conductive effects as well as the structural behavior. The resulting knowledge increases our understanding of the complex, underlying processes, ultimately aiding the design of new materials in which the desired chemical or physical properties are enhanced.

2.2 X-Ray Diffraction

The field of crystallography developed to great heights after the discovery of X-rays by Wilhelm Conrad Röntgen in 1895 (Novelline and Squire 2004). The first X-ray experiment was performed by Max von Laue in Munich in 1912 (Purrington 1997). At the time of von Laue's work, there was some experimental evidence that X-rays were high energy particles; other data indicated that X-rays might be waves. Von Laue surmised that, if X-rays were waves, they would have rather short wavelengths (on the order of 1×10^{-10} m) and the dimensions of the objects in crystals would be the appropriate size to produce the phenomenon of diffraction. He exposed a crystal of copper sulfate to X-rays and recorded the diffraction pattern on a piece of photographic film, the experimental setup similar to that shown in Fig. 2.1.

This experiment proved the wave nature of X-rays and began the exploration of molecular structure by X-ray diffraction methods. Modern X-ray data collection is substantially the same as von Laue's experiment though various kinds of electronic detectors are used rather than photographic film.

An X-ray photo contains two important pieces of information about the crystal structure, the shape and size of the unit cell and the distribution of electron density throughout the unit cell. A useful explanation of von Laue's experiment was formulated by Bragg (1913) in Cambridge, in 1912. Bragg considered the diffractions to arise from 'reflections' of separate X-ray waves from parallel planes of electron density.

2.2.1 Bragg's Equation

X-ray diffraction results from the interaction between X-rays and electrons of atoms. Depending on the atomic arrangement, interfaces between the scattered rays are constructive when the path difference between two diffracted rays differs by an integral number of wavelengths. This selective condition is described by Bragg's equation also called "Bragg's law" (Bragg 1912).

Crystals consist of a periodic arrangement of atoms or molecules that form a crystal lattice. In such an arrangement of atoms numerous planes run in different directions through the lattice points (atoms, molecules), horizontally, vertically, and diagonally which are the lattice planes. All the planes parallel to a lattice plane are also lattice planes and are set at a distance apart from each other which is

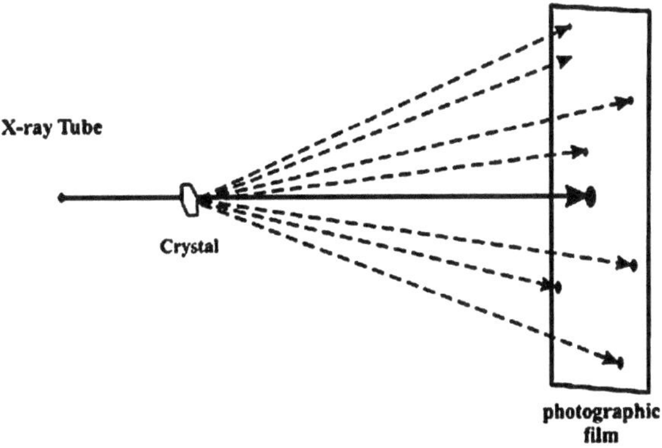

Fig. 2.1 X-rays passing through a crystal and falling on a photographic plate

called the lattice plane distance 'd'. Bragg diffraction occurs when electromagnetic radiation or subatomic particle waves with wavelength comparable to atomic spacings are incident upon a crystalline sample, scattered in a specular fashion by the atoms in the system, undergo constructive interference. When the scattered waves interfere constructively, they remain in phase since the path length of each wave is equal to an integer multiple of the wavelength. The path difference between two waves undergoing constructive interference is given by 2d Sin θ, where θ is the scattering angle. This leads to Bragg's law which describes the condition for constructive interference from successive crystallographic planes (*h,k,l*) of the crystalline lattice (Azaroff 1968).

In Fig. 2.2, two monochromatic X-ray beams of wavelength 'λ' are represented by wave 1 and wave 2. The spacing between the atomic planes occurs over the distance, d. Wave 2 reflects off at the upper atomic plane at an angle θ equal to its angle of incidence. Similarly, wave 1 reflects off at the lower atomic plane at the same angle θ. although wave 1 is in the crystal, it travels a distance of 2d farther than wave 2. Hence the path difference between waves 1 and 2 is 2d. If this distance 2d is equal to an integral number of wavelengths nλ, then waves 1 and 2 will be in phase on their exit from the crystal and constructive interference will occur. If the distance 2d is not an integral number of wavelengths, then destructive interference will occur and the waves will not be as strong as when they entered the crystal. Thus, the condition for constructive interference to occur is nλ = 2d.

$$\begin{aligned}
&\text{In figure, } BC = d \sin \theta, DD = d \sin \theta \\
&\text{The path difference} = BC + BD \\
&BC + BD = 2d \sin \theta \\
&\text{Thus, } n\lambda = 2d \sin \theta
\end{aligned} \tag{2.1}$$

This is known as Bragg's Law (Bragg 1913) for X-ray diffraction.

On the basis of Bragg's Law, by measuring the angle 'θ', the wavelength 'λ', the chemical elements can be determined, if the lattice plane distance 'd' is known, or, if the wavelength 'λ' is known, the lattice plane distance 'd' and thus the crystalline structure can be determined. This provides the basis for two measuring

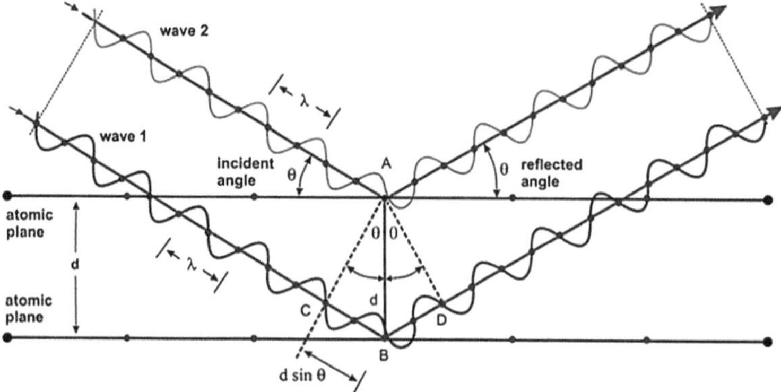

Fig. 2.2 Schematic representation of X-ray reflection from two parallel planes of a crystal

techniques for the quantitative and qualitative determination of chemical elements and crystalline structures, depending on whether the wavelength 'λ' or the 2d-value is identified by measuring the angle 'θ' (George 2006).

2.2.2 Electron Density

According to the theory of quantum mechanics an electron's position can only be described statistically. The probability of finding an electron at one point or another can be calculated. This calculation produces a quantity called electron density, a number that tells us the relative probability of finding an electron at a particular point in space. Quantum mechanics also says that an electron can be viewed as a stationary wave, or, as a cloud of negative charges. The electron density is a periodic function of position in the crystal, rising to a maximum at the point where an atom is located and dropping to a low value in the region between atoms (Cullity and Stock 2001).

An electron's wave characteristics are described mathematically by its orbital or wave function, ψ. A wave function assigns a number to each point in space, and the numbers oscillate so that they are positive at some locations and negative at others. The electron density function, ρ, is equal to Ψ^2. This guarantees that ρ will always have a positive value, and this value expresses the relative probability of finding an electron at a particular location.

$$\Psi \qquad \leftrightarrow \qquad \Psi^2 = \rho \qquad (2.2)$$

$$\text{wave function} \qquad\qquad \text{probability}$$
$$\text{(probability amplitude)} \qquad \text{(or electron density)}$$

A 'particle' electron can also be described by a 'wave' function or orbital. The orbital, whether atomic or molecular, covers a region of space and does not move. A moving electron looks like a stationary cloud of charges. The physical interpretation of the electron density function $\rho(r)$ is that ρ dr is the probability of finding an electron in a volume element dr, i.e., electron density in this volume.

2.2.3 Structure Factor

The crystal structure, which is determined by the lattice parameters and the atomic positions within the unit cell, is an essential element for the diffraction intensities. Each atom scatters according to its electronic distribution following dependence on $\sin\theta/\lambda$. When there is more than one atom per unit cell, the interference between the waves scattered by each atom is considered as the diffraction unit. The addition of the waves scattered by each of the atoms of the unit cell, considering that there is coherence in time and space is called the structure factor (Marin and Dieguez 1999).

The structure factor F_{hkl} is the resultant of j waves scattered in the direction of the reflection hkl by the j atoms in the unit cell (Stout and Jensen 1968).

The structure factor may be expressed as

$$F_{hkl} = F_{hkl}\exp(i\alpha_{hkl}) = \sum_j f_j \exp\left[2\pi i\left(hx_j + ky_j + lz_j\right)\right] \tag{2.3}$$

$$F_{hkl} = \sum_j f_j \cos\left[2\pi\left(hx_j + ky_j + lz_j\right)\right] + i\sum_j f_j \sin\left[2\pi\left(hx_j + ky_j + lz_j\right)\right] \tag{2.4}$$

$$F_{hkl} = A_{hkl} + iB_{hkl} \tag{2.5}$$

where the sum is over all atoms in the unit cell, x_j, y_j, z_j are the positional coordinates of the jth atom, f_j is the scattering factor of the jth atom, and α_{hkl} is the phase of the diffracted beam.

The scattering factor of the atom is,

$$f_j = \frac{amplitude\ of\ the\ radiation\ scattered\ from\ the\ atom}{amplitude\ of\ the\ radiation\ scattered\ from\ a\ single\ electron} \tag{2.6}$$

The structure factor describes the way in which an incident beam is scattered by the atoms of a crystal unit cell, taking into account the different scattering power of the elements through the term f_j. Since the atoms are spatially distributed in the unit cell, there will be a difference in phase when considering the scattered amplitude from two atoms. This phase shift is taken into account by the complex exponential term. The atomic form factor or scattering power, of an element depends on the type of radiation considered as electrons interact with matter through different processes.

If the summation over discrete atoms in the following structure factor expression

$$F_{hkl} = \sum f_j e^{2\pi i \left(hx_j + ky_j + lz_j \right)} \tag{2.7}$$

is replaced by an integration of a continuous, cyclic electron density function, ρ, an expression is obtained that is of the form of a Fourier Transform

$$F_{hkl} = \int_V \rho(x.y.z) e^{2\pi i \left(hx_j + ky_j + lz_j \right)} \, dV \tag{2.8}$$

This implies that the electron density is the Fourier Transform of the Structure Factor. Likewise, the Structure Factor is the Fourier Transform of the electron density.

$$\rho(x.y.z) = \int_V F_{hkl} e^{-2\pi i (hx + ky + lz)} \, dV \tag{2.9}$$

While this equation is a useful depiction of the relationship between electron density and the structure factor, it is usually used in the form of a summation

$$\rho(x, y, z) = \frac{1}{V} \sum_h \sum_k \sum_l F_{hkl} e^{-2\pi i (hx + ky + lz)} \tag{2.10}$$

According to Eq. 2.10 the electron density, ρ, can be calculated at any point (x, y, z) by constructing a Fourier Series which has coefficients that are equal to the Structure Factors (Warren 1990). This is the basic equation of crystallography.

It enables the calculation of a three-dimensional electron density map throughout the unit cell. Maxima in the electron density map define the locations of individual atoms.

2.3 Crystal Structure Determination from Diffraction Data

Structure determination in crystallography refers to the process of elaborating the three-dimensional positional coordinates (and also, usually, the three-dimensional anisotropic displacement parameters) of the scattering centers in an ordered crystal lattice. Crystal structure determination from diffraction data involves unit cell determination, structure solution, and structure refinement.

The aim of structure solution is to obtain an approximate description of the structure, using the unit cell and space group determined, but starting from no knowledge of the arrangement of atoms or molecules within the unit cell. Typical mineral structures contain several thousand unique reflections, whose spatial arrangement is referred to as a diffraction pattern. Indices (hkl) may be assigned to each reflection, indicating its position within the diffraction pattern. This pattern has a reciprocal Fourier transform relationship with the crystalline lattice and the unit cell in real space. This step is referred to as the solution of the crystal structure. If the structure solution is a sufficiently good approximation to the true structure, then a good quality structure can be obtained by structure refinement.

2.3.1 Structure Refinement

Once a structure solution has been achieved, there are actually two structure models, a calculated model based on the approximate co-ordinates obtained from interpretation of a three-dimensional electron density map,

$$F_{hkl} = \sum f_j e^{2\pi i \left(h x_j + k y_j + l z_j \right)}$$

(2.11)

and an observed model based on the calculation of structure factors from experimental intensities,

$$|F_{hkl}| = K \sqrt{I_{hkl}}$$

(2.12)

The task now is to adjust the various atomic parameters so that the calculated structure factors match the observed structure factors as closely as possible. One of the ways to measure the agreement between the observed and calculated models is with the Residual Index (more commonly referred to as the R-factor)

$$R = \frac{\sum |F_{\mathrm{obs}} - F_{\mathrm{calc}}|}{\sum |F_{\mathrm{obs}}|}$$

(2.13)

In addition to the approximate atomic x, y, z-coordinates, there is another factor influencing the magnitudes of the structure factors, the thermal motion. Although the structure determination is done on the solid phase, the atoms still have some thermal motion (vibration and rotation). This thermal effect is introduced into the structure factor equation by a factor that serves to attenuate the atomic scattering factor,

$$f_j = f_o e^{-B \frac{\left(\sin^2 \theta \right)}{\lambda^2}}$$

(2.14)

where the temperature factor (Warren 1990) B, is related to the mean-square amplitude of vibration,

$$B = 8\pi^2 \overline{u^2}$$

(2.15)

As B increases, the scattering power of atom $j(f_j)$ decreases. To represent anisotropic thermal motion, a total of six parameters per atom are necessary. Three of these parameters provide the orientations of the principal axes of the ellipsoid produced by anisotropic thermal motion; the other three parameters represent the magnitudes of displacement along the ellipsoid axes.

With thermal effects included, the complete expression for the calculated structure factor is

$$F_{hkl} = \sum f_j e^{2\pi i \left(hx_j + ky_j + lz_j\right)} e^{-B\frac{\left(\sin^2 \theta\right)}{\lambda^2}} \tag{2.16}$$

In order to refine the structure, the coordinates and temperature factors must be adjusted so that these calculated F_{hkl}'s match, as closely as possible, the observed F_{hkl}'s derived from the experimentally measured intensities.

2.3.2 Theoretical Models in Structure Analysis

The classical approach to solve the structure of a material is to build a physical model that is consistent with all the information known about the material. An obvious way to expedite the model building process would be to exploit the power of a computer, and several programs have been developed to achieve the accurate structure. The validity of a structural model is that it accounts for all the data available. The calculated powder diffraction pattern should not only match the observed data but should be consistent with spectroscopic data and with the physical properties of the material. Some of the versatile methods such as Rietveld method (Rietveld 1969), Maximum Entropy method (Collins 1982), Multipole method (Hansen and Coppens 1978), and Pair distribution function (Proffen and Billinge 1999) have been followed in this study to analyze the structural properties of metals and alloys.

2.4 Methods in X-Ray Crystallography

In terms of the specimen handled, two methods can be identified, viz., the single crystal method and the powder method.

1. Single-crystal X-ray diffraction
2. Powder X-ray diffraction

The single crystal diffraction technique, using relatively large crystals of the material, gives a set of separate data from which the structure can be obtained. However, most materials of technical interest cannot grow large crystals, so one has to resort to the powder diffraction technique using material in the form of very small crystallites. In reality, the two techniques are highly complementary, have their own strengths and weaknesses and domains of applicability, and one will never supercede the other. However, the domain where powder diffraction is having an impact is certainly growing and diversifying. In general, the structure solution process poses more significant challenges, which has prompted much research in recent years on the development of new strategies and techniques in this field.

2.4.1 Structure Determination from Single-Crystal X-Ray Diffraction

Within the realm of crystallography, single-crystal X-ray diffraction is by far the most commonly used technique available for the determination of the crystal and molecular structure of crystalline solids (Goeta and Howard 2004). Single-crystal X-ray Diffraction is a non-destructive analytical technique which provides detailed information about the internal lattice of crystalline substances, including unit cell dimensions, bond-lengths, bond-angles, and details of site-ordering. In the single-crystal method, the whole specimen is a single piece, without any discontinuity in the lattice arrangements. Typical mineral structures contain several thousand unique reflections, whose spatial arrangement is referred to as a diffraction pattern. This pattern has a reciprocal Fourier transform relationship with the crystalline lattice and the unit cell in real space (Shmueli 2008).

The measured intensities I_{hkl} (corresponding to scattering from a lattice plane with Miller indices h, k, l) are reduced to structure amplitudes F_{hkl} by the application of a number of experimental corrections

$$F_{hkl}^2 = I_{hkl}\left(KL_pA\right)^{-1} \tag{2.17}$$

where k is a scale factor, L_p the Lorentz–polarization correction, and A the transmission factor representing the absorption of X-rays by the crystal.

Once the initial crystal structure is solved, an iterative refinement procedure is essential to attain the best possible fit between the observed and calculated crystal structure. The most common approach is to perform a least-squares minimization between the experimental structure factors and those calculated by varying the adjustable parameters of the structural model. These normally include atomic positions, anisotropic displacement parameters, occupancies, chemical bond lengths and angles, and other geometric characteristics of a molecule.

The results of the structure refinement yield a list of atom X–Y–Z assignments in the unit cell, shape of the anisotropic intensity center for each atom (thermal parameters), and the distance of the nearest atomic neighbors. The quality of a solution is assessed by the values of R, wR, and GooF.

The final structure solution will be presented with the residual R factor, which gives the percent variation between the calculated and observed structures.

$$R = \frac{\sum |F_{\mathrm{obs}} - F_{\mathrm{calc}}|}{\sum |F_{\mathrm{obs}}|} \tag{2.18}$$

The R-value is used to indicate improvements or reductions in the quality of fit between model and observation. In the expression above F_{obs} and F_{calc} are the observed and calculated structure amplitudes and the deviations are summed over all experimentally recorded intensities. Ideal solutions would have R-value of 0, however, due to random errors, this is never achieved. R-values (listed as percents) of less than 5% are considered good solutions, high quality samples will often

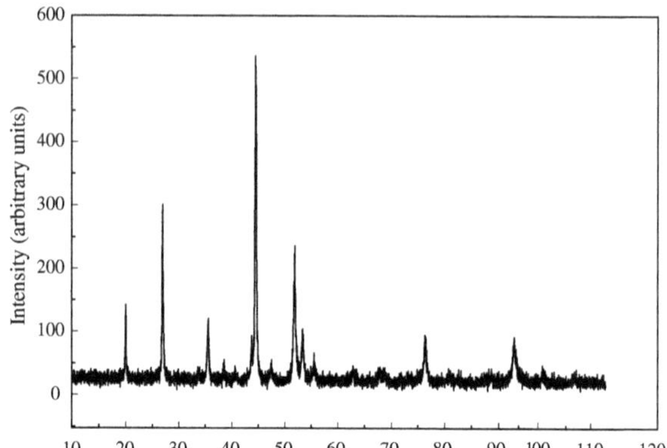

Fig. 2.3 A typical powder diffraction profile

result in R-values lower than 2.5%. wR refers to squared F-values. GooF refers to the "goodness of fit" of the solution. In addition to the difference in F values, the GooF also takes into account the number of observed reflections and the parameters used. At the end of refinement, the GooF should approach 1.

2.4.2 Powder Diffraction

The possibility of using powder diffraction methods to study materials was recognized shortly after the discovery of X-ray diffraction by Laue and Von Knipping in 1910. Although, the powder method was developed as early as 1916 by Debye and Scherrer, for more than 50 years its use was almost exclusively limited to qualitative and semi-quantitative phase analysis and macroscopic stress measurements. In 1922, Bain using the powder photograph method of Debye and Hull carried out some very important pioneering work (Weiss 1925). Within a few years many others including Bragg and Pauling, had exploited the powder method to study a wide range of materials, including metals, minerals, and simple organic solids. Quantitative analysis of the pattern using modern computers and software yields a wealth of additional information about the sample structure. Figure 2.3 shows a typical powder diffraction profile.

Powder diffraction has played a central role in structural physics, chemistry, and materials science over the past 20 years. Important advances in structural studies of materials ranging from high temperature superconductors and fullerenes to zeolites and high-pressure research have relied heavily on the powder diffraction technique.

Although most of the structures determined from powder diffraction data have been solved in the last few years, notable contributions have been from Zachariasen (1948) and Berg and Werner (1977). Zachariasen, in particular, used a number of ingenious methods to solve crystal structures from powders. His work on β-plutonium (Zachariasen and Ellinger 1963), for example, utilized differential thermal expansion to resolve Bragg peak overlap. Many of the early zeolite structures that were solved from powder diffraction data involved model building and significant chemical intuition (Breck et al. 1956; Kokatalio et al. 1978), and these concepts are now implemented in computer algorithms.

Solving a structure from powder diffraction data has developed rapidly over the last 10 years. Prior to 1990, very few unknown crystal structures had been determined directly from powder diffraction data. Today, numerous crystal structures, both organic and inorganic, have been solved from powder data. Developments in instrumentation, computer technology, and powder diffraction methodology have all contributed to this increased success rate.

2.4.2.1 Structure Determination from Powder Diffraction

The first choice is the selection of radiation source and instrumentation geometry. The first step in data analysis is indexing of the diffraction pattern. From the unit cell and space group the diffraction intensity associated with each reflection should be determined by applying whole-profile fitting. The intensity extraction can be performed using either a least-squares method (Pawley 1981) or an iterative approach (Le Bail et al. 1988).

Rietveld proposed a simple summation approach to the evaluation of an observed structure factor magnitude for partially and completely overlapped reflections, if the calculated diffraction pattern is good but the observed and calculated structure factor magnitudes are different from one another.

The peak area is proportional to the square of the structure factor magnitude and the problem reduces to finding the peak area. For an isolated peak, the observed peak area is easy to evaluate. For overlapping peaks, the contribution for a given reflection weighted by the calculated peak contribution for that reflection is divided by the sum of the calculated peak values for each contributing reflection. The estimated standard deviations of observed intensities are not normally included as part of the standard Le Bail approach, they may nevertheless be evaluated.

Pawley (1981) declared a method for determining Bragg peak intensities from powder diffraction data in the absence of structural model. The variables associated with peak positions and widths are the same, the variables associated with peak areas are peak areas themselves. Although the Pawley method was introduced some 6 years earlier than its counterpart, LeBail method is currently still the more popular approach.

The final step in the structure determination maze is the completion of the structure and the refinement of the structural parameters using the Rietveld method (Rietveld 1969). The structural proposal from the structure determination can be confirmed only when the refinement has been brought to a successful conclusion. Chemical information and intuition play important roles in guiding the path in the maze.

2.5 The Rietveld Method

The drawback of the conventional powder method is that the data grossly overlap, thereby preventing proper determination of the structure. The major breakthrough in the value of the powder method as a quantitative tool was the development of the Rietveld method (Rietveld 1969), a technique for crystal structure refinement which, for the first time, made use of the entire powder pattern instead of analyzing individual, non-overlapped, Bragg reflections separately. This approach minimizes the impact of overlapped and degenerate peaks by calculating the entire powder pattern of a crystalline model, including various experimental and sample dependent peak-broadening effects. Parameters in the model such as atomic positions, lattice parameters, and experimental factors that affect peak-shape and background are varied, using a least-squares approach, until the agreement between the calculated and measured diffraction profiles are optimized and the model is refined by iterative procedure.

The method was first reported for the diffraction of monochromatic neutrons where the reflection-position is reported in terms of the Bragg angle 2θ. The method was quickly extended from reactor neutron data, with its nice Gaussian line-profiles and lack of atomic form-factor, to in-house X-ray powder diffraction, synchrotron powder diffraction, and time-of-flight neutron data from pulsed spallation sources, and to refinements of incommensurate and magnetic structures. The application of the Rietveld method to neutron data in the early 1970s was soon followed by its extension to laboratory X-ray diffractometer data (Malmros and Thomas 1977; Young et al. 1977). The technique has been applied to a wide range of solid-state problems and has been reviewed by several authors during the last 25 years (Cheetham and Taylor 1977; Hewat 1986).

The problem of the more complex peak shape was resolved by employing alternative peak-shape functions, such as the Lorentzian and the pseudo-Voigt. Other problems that can plague X-ray studies include preferred orientation and poor powder averaging (graininess), both of which arise from the fact that X-rays probe a smaller sample volume than neutrons; these were addressed by paying closer attention to the data collection strategy. The accuracy and precision of a structure refinement from X-ray data can normally be optimized by collecting high-resolution data at a synchrotron source (Cox et al. 1983). The resolution of the powder diffractometers at second and third generation sources is so good that sample imperfections now play a major role in determining the shape of the Bragg peaks.

2.5.1 The Rietveld Strategy

Powder diffraction patterns are collected in a step scan mode. Intensity and two-theta axes of Bragg diffraction are collected for the crystal structure determination (Young 1995). The only wavelength and technique independent scale is in reciprocal space units or momentum transfer Q, which is historically rarely used in powder diffraction but very common in all other diffraction and optics techniques. The relation is

$$Q = \frac{4\pi \sin(\theta)}{\lambda} \tag{2.19}$$

2.5.1.1 Peak Shape

The shape of a powder diffraction reflection is influenced by the characteristics of the beam, the experimental arrangement, and the sample size and shape (Rietveld 1969). In the case of monochromatic neutron sources the convolution of the various effects has been found to result in a reflex almost exactly Gaussian in shape. If this distribution is assumed then the contribution of a given reflection to the profile y_i at position $2\theta_i$ is:

$$y_i = I_K \exp\left[\frac{4\ln(2)}{H_k^2}(\theta_i - \theta_k)^2\right] \tag{2.20}$$

where H_k is the full width at half peak height (full-width half-maximum), $2\theta_k$ is the center of the reflex, and I_k is the calculated intensity of the reflex (determined from the structure factor, the Lorentz factor, and multiplicity of the reflection).

At very low diffraction angles the reflections may acquire an asymmetry due to the vertical divergence of the beam. Rietveld used a semi-empirical correction factor, A_s to account for this asymmetry.

$$A_S = 1 - \left[\frac{sP(2\theta_i - \theta_k)^2}{\tan\theta_k}\right] \tag{2.21}$$

where P is the asymmetry factor and s is +1, 0, -1 depending on the difference $2\theta_i - 2\theta_k$ being positive, zero, or negative respectively. At a given position more than one diffraction peak may contribute to the profile. The intensity is simply the sum of all reflections contributing at the point $2\theta_i$.

2.5.1.2 Peak Width

The width of the diffraction peaks are found to broaden at higher Bragg angles. This angular dependency (Caglioti et al. 1958) was originally represented by

$$H_K^2 = U \tan^2\theta_k + V \tan^2\theta_k + W \tag{2.22}$$

where U, V, and W are the half-width parameters and may be refined during the fit.

2.5.1.3 Preferred Orientation

In powder samples there is a tendency for plate-like or rod-like crystallites to align themselves along the axis of a cylindrical sample holder. In solid polycrystalline samples the production of the material may result in greater volume fraction of certain crystal orientations (commonly referred to as texture). In such cases the reflex intensities will vary from that predicted for a completely random distribution. Rietveld allowed for moderate cases of the former by introducing a correction factor:

$$I_{\text{corr}} = I_{\text{obs}} \exp\left(-G\alpha^2\right) \tag{2.23}$$

where I_{obs} is the intensity expected for a random sample, G is the preferred orientation parameter, and α is the acute angle between the scattering vector and the normal of the crystallites.

2.5.1.4 Refinement

The principle of the Rietveld Method is to minimise a function M which analyzes the difference between a calculated profile y(calc) and the observed data y(obs). Rietveld (1969) defined such an equation as:

$$M = \sum_i W_i \left\{ y_i^{\text{obs}} - \frac{1}{c} y_i^{\text{calc}} \right\}^2 \tag{2.24}$$

where W_i is the statistical weight and c is an overall scale factor such that $y^{\text{calc}} = cy^{\text{obs}}$.

Rietveld (1969) method calculates the entire powder pattern using a variety of refinable parameters and improves the selection of these parameters by minimizing the weighted sum of the squared differences between the observed and the calculated pattern using least square methods. Thus the systemic and accidental peak overlap is overcome.

The method has been so successful that nowadays the structure of materials, in the form of powders, is routinely being determined, nearly as accurately as the results obtained by single-crystal diffraction techniques. An even more widely used application of the method is in determining the components of chemical mixtures. This quantitative phase analysis is now routinely used in industries ranging from cement factories to the oil industry. The success of the method can be gauged by the publication of more than a thousand scientific papers yearly using it.

2.5.2 Rietveld Refinement

The Rietveld refinement was performed using the softwares JANA 2000 (Petříček et. al. 2006) and an improvised version JANA 2006 (Petříček et. al. 2006). JANA

2000 and JANA 2006 are crystallographic programs for structure analysis of crystals periodic in three or more dimensions from diffraction data. It is focused to solution, refinement, and interpretation of difficult, especially modulated structures. It calculates structures having up to three modulation vectors from powder as well as single-crystal data measured with X-ray or neutron diffraction. The input diffraction data can be unlimitedly combined, the combination of powder neutron data with single-crystal X-ray data being a typical example. The structure solution can be done using the builtin charge flipping algorithm or by calling an external direct methods program. Jana can handle multiphase structures (for both powder and single-crystal data), twins with partial overlap of diffraction spots, commensurate, and composite structures. It contains powerful transformation tools for symmetry (groupsubgroup relations), cell parameters, and commensurate—super cell relations. A wide scale of constrains and restrains is available including a powerful rigid body approach and the possibility to define a local symmetry affecting only part of the structure.

In the refinement using the software, the observed profiles are matched with the profiles constructed similarly by using pseudo-voigt (Wertheim 1974) profile shape function of Thompson, Cox and Hastings (1987) modified to some extent that accommodate various Gaussian FWHM parameters and Scherer coefficient for Gaussian broadening. The profile asymmetry is introduced by employing multi-term Simpson rule integration devised by Howard (1982) that incorporates symmetric profile shape function with different coefficient for weights and peak shift. Jana 2006 also employs the correction for preferred orientation which is independent of diffraction geometry using March—Dollase function (March 1932; Dollase 1986). The calculated profiles thus evolved are compared with the observed ones. Finally, the structure factors evolved from the Rietveld refinements were further utilized for the estimation of charge density in the unit cell.

2.6 Multipole Method

The distribution of positive and negative charges in a crystal fully defines physical properties like the electrostatic potential and its derivatives, the electric field, and the gradient of the electric field. The electrostatic potential is of importance in the study of intermolecular interactions, and has received considerable attention during the past two decades. It plays a key role in the process of molecular recognition, including drug-receptor interactions, and is an important function in the evaluation of the lattice energy. The evaluation of the electrostatic potential and its derivatives may be achieved directly from the structure factors, or indirectly from the experimental electron density by the multipole formalism.

The ability to measure the experimental charge distribution in crystals from the intensities of the scattered X-rays was realized almost immediately after the discovery of X-ray diffraction. Notwithstanding this early recognition, the technical developments of the 1960s and beyond, which occurred in difftractometry, automation of data collection, low temperature techniques, computers were needed to achieve a breakthrough in the method. The accurate crystallographic methods developed during the past decades led not only to a much better precision in atomic coordinates, but also to crucial information about the charge distribution in crystals. This experimentally obtained distribution can be compared directly with theoretical results, and can be used to derive other physical properties such as electrostatic moments, the electrostatic potential, and lattice energies, which are accessible by spectroscopic and thermodynamic measurements. This broad interface with other physical sciences is one of the most appealing aspects of the field.

Recently, developments in theoretical methods have facilitated the calculation of charge densities of sufficient accuracy so that they can be compared with the experimental charge distributions. However, these are limited for molecular crystals consisting of only light atoms. Such comparisons would reveal the deficiencies in both theory and experiment. The state of art now has reached a stage where it is possible to derive properties like net charges, molecular dipole moments and electrostatic potentials and to fit atom-centered spherical harmonic functions (Guru Row 1983).

The charge density effects in Beryllium metal were measured (Larsen et al. 1980), while the absolute scale was established with an Sm source (Hansen et al. 1987). The bonding between Chromium atoms have been studied and analyzed as dichromium tetracarboxylates have metal–metal bond lengths which vary by as much as 0.7 Å (Cotton and Stanley 1977). To explain the properties of β'NiAl (Fox and Tabernor 1991) measured four low angle structure factors which showed a depletion of density around both Ni and Al atoms, and a buildup of about 0.13 $e\text{Å}^{-3}$ along the (111) direction halfway between Ni and Al nearest neighbors.

2.6.1 Multipole Electron Density Model

Expansion of the charge density in a crystal in terms of a series of nucleus-centered spherical harmonic functions was first applied by De Marco and Weiss (1965) and significantly complemented by Dawson (1967) and collaborators in their study of bonding in diamond-type structures. In the earliest treatment the radial function of the valence shell is that of the isolated atom, but the Dawson formalism allows a modified Gaussian radial function to have the same radial dependence as the isolated atom. A more flexible radial function consisting of a set of harmonic oscillator function wave functions was used by Kurki-Suonio (1968). Least squares adjustment of the parameters in the Dawson model was carried out by McConnell and Sanger (1970), while Stewart's functions include a full set of spherical harmonics truncated at a level chosen by practical considerations such as adequacy in describing the aspherical deformations and available computing facilities. A related model of $\cos^n$ type, which is linear combinations of spherical harmonics, was developed (Hirshfeld 1971; Harel and Hirshfeld 1975) and applied to a number of organic molecules.

Such models have the obvious merit of providing an analytical description of the charge density and may be used in the calculation of physical properties based on charge density distribution (Stewart 1972). As their use should increase as more accurate data sets become available a critical analysis of the results has been undertaken.

This analysis requires X-ray data sets specifically collected for charge density studies.

2.6.2 Mathematical Approach of Multipole Electron Density Model

The density model (Hansen and Coppens 1978) consists of a superposition of harmonically vibrating aspherical atomic density distributions:

$$\rho(r) = \sum_{k}^{\text{atoms}} \rho(r - r_k - u) \otimes t_k(u) \tag{2.25}$$

where $t_k(u)$ is a Gaussian thermal displacement and $\otimes$ indicates a convolution. Each atomic density is described as a series expansion in real harmonic functions through fourth–order (y_{lm}).

$$\rho_{\text{atomic}}(r) = P_c\rho_{\text{core}} + P_v k'^3 \rho_{\text{valence}}(k'r) + \sum_{l=0}^{4} -''^3 R_l(k''r) \sum_{m=-1}^{l} P_{lm} y_{lm}\left(\frac{\vec{r}}{r}\right) \tag{2.26}$$

Here P_c, P_v, and P_{lm} are population coefficients. The total number of electrons associated with one atom is equal to $P_c + P_v + P_{00}$, since the higher terms with $l \neq 0$ integrate to zero when integration is performed over all space. The functions ρ_{core} and ρ_{valence} are chosen as the Hartree–Fock (HF) densities of the free atoms normalized to one electron, but the valence function is allowed to expand and contract by adjustment of the variable radial parameter k' (Coppens et al. 1979). ρ_{core} and ρ_{valence} are constructed from the canonical Hartree–Fock atomic orbitals.

The radial functions $R_l(r)$ of the other terms are

$R_l(r) = \frac{\zeta^{n_l+3}}{(n_l+2)!} r^n l \exp(-\zeta_l r)$, where in principle n_l can take any positive integer value. Starting values of the orbital exponent ζ are modified by a variable parameter k'', such that $\zeta' = k''\zeta$.

The spherical harmonic functions y_{lm} and the normalization factors based on

$\int |P_{lm}| d\tau = 2.$ This normalization implies that $P_{lm} = 1$ one electron has been moved from the negative to the positive lobes of the deformation functions with $P \geq 1$.

The structure factor corresponding to (2) and (3) becomes:

$$F(h) =$$

$$\sum_{k}^{\text{atoms}} \sum_{p}^{\text{symmetry}} \left[P_c f_{\text{core}}(h) + P_v f_{\text{valence}}\left(\frac{h}{k''}\right) + \sum_{l=0}^{4} \varphi kl\left(\frac{h}{k''}\right) \sum_{m=-1}^{l} P_{klm} y_{klm}\left(\frac{h_p}{h}\right) \right] \times$$

$$\exp 2\pi i h.r_{kp} T_k(h)$$

$$(2.27)$$

where f_{core} and f_{valence} are the Fourier transforms of ρ_{core} and ρ_{valence} respectively, and φ_{kl} is the Fourier transform of R_{kl} defined as:

$$\varphi_{kl}(h) = 4\pi i^l \int_0^\infty R_{kl}(r) j_l(2\pi hr)\, r^2 dr \qquad (2.28)$$

The temperature factor T_k is the Fourier transform of t_k and h_p and r_{lp} are symmetry transformed scattering and position vectors. We note that the model reverts to the free-atom model when $\kappa' = 1$ and the $P_{klm'}$ s are zero. It is in this respect, similar to Hirshfeld's deformation model (Harel and Hirshfeld 1975), but different from the single Slater-type models used by a series of authors (Cromer et.al. 1976; Price and Maslen 1978; Chen et.al. 1977).

2.6.3 Criteria for Judging Aspherical Atom Refinements

Although an aspherical atom model (multipole) refinement based on Eq. 2.25 may lead to a lowering of agreement indices, the improvement is not necessarily significant because of the large number of parameters involved, and the results are not necessarily physically meaningful as they may be affected by parameter correlation or systematic errors in the measurements. The following tests of the multipole refinements have therefore been applied.

2.6.3.1 Significance of Improvement of the Fit Between Calculated and Observed Structure Factors

For an adequate model the goodness of fit g, defined as

$$g = \left(w_i \Delta_i^2 / v \right)^{\frac{1}{2}} \qquad (2.29)$$

should tend to 1 as refinement proceeds.

For a refinement of F, Δ equals $|F_{\text{obs}}| - |kF_{\text{calc}}|$, w_l represents the weight assigned to each of the reflections from an estimated of experimental accuracy and v is the number of degrees of freedom (i.e., the number of independent observations n, minus the number of independently varied parameters p).

In comparison of different models we use the error function

$$\in = vg^2 = \sum w_i \Delta_i^2 \tag{2.30}$$

which follows an χ^2 distribution with v degrees of freedom for a fully refined model. The error function can be tested by a modification of the R-factor test:

$$\frac{\varepsilon' - \beta}{\varepsilon} - \frac{d}{n-p} F_{d,n-p} \tag{2.31}$$

Or

$$\frac{\varepsilon'}{\varepsilon} = 1 + \frac{d}{n-p} F_{n-p} \tag{2.32}$$

where d is the dimensionality of the hypothesis, and the tabulated F distribution is the ratio of two χ^2distributions.

2.6.3.2 Residual Density Maps

A residual density map, from a Fourier summation based on $\left(\frac{F_{obs}}{k} - |F_{calc}|\right)$ as is commonly used in structure analysis will show to what extent the multipole expansion has been successful in describing the features of the density distribution. It does not contain information, however, about the significance of residual features. For this purpose the residual density distribution may be compared with a map representing the position dependence of the estimated standard deviation in ρ_{obs} as described by Rees (1976, 1977) and calculated in several previous studies (Stevens and Coppens 1976; Coppens et.al. 1977).

2.6.3.3 Parameter Bias

A successful refinement should give parameters with minimal bias due to atomic asphericity. The rationale for this requirement is twofold. First, deformation maps (i,e., total minus 'sum of spherical atom' densities) are quite sensitive to errors in positional and thermal parameters, and second, physical properties to be calculated from the refinement results, such as dipole and quadrupole moments, depend on the nuclear positions as well as charge distribution.

2.6.3.4 Derived Properties

One of the justifications for performing the multipole refinement is the experimental determination of derived physical properties (Stewart 1972). Properties such as net molecular charges, dipoles, and quadrupole moments may also be obtained by direct-space integration of the experimental charge distribution.

2.6.4 Multipole Refinement Strategy

The raw data is refined for their structural parameters with extinction, absorption, and TDS corrections. Structure factors are refined to yield the smallest possible reliability indices. The results are compared for the thermal parameters obtained using MEM. The refined structure factors are further refined for the population and k-parameters using the formalism proposed by Hansen and Coppens (1978). In this refinement, the canonical Hartree–Fock atomic orbitals of the free atoms normalized to one electron are used for the construction of ρ_{core} and $\rho_{valence}$ charge densities.

2.6.4.1 Static and Dynamic Deformation Density

The multipole charge density is usually analyzed by means of deformation density maps. The multipole density consists of superposition of harmonically vibrating aspherical atomic density distribution as

$$\rho(r) = \sum_{k}^{\text{atoms}} \rho_k(r - r_k - u) \otimes t_k(u) \tag{2.33}$$

where $t_k(u)$ is a Gaussian thermal-displacement and $\otimes$ indicates a convolution. When the thermal contribution is de-convoluted, this density becomes the density of a static atom. The deformation between a static model density and the observed density reveals the contribution from thermal displacement of charge distribution. Thus

$$\Delta\rho_{\text{deformation}} = \rho_{\text{multipole}} - \rho_{\text{observed}} \tag{2.34}$$

gives static deformation density when thermal contribution is de-convoluted and dynamic deformation density when it is convoluted with the radial dependent multipole charge density function.

2.6.5 Significance of Multipole Model

It is appropriate to take a critical look at the aspherical-atom multipole (pseudo atom) model, as expressed in a number of algorithms (Hirshfeld 1971, 1977a; Stewart 1976), including the Hansen and Coppens model (1978). The pseudo atom model has significant advantages and its introduction has greatly contributed to the increasing application of experimental results in charge-density analysis,

1. Experimental noise is generally not fitted by the model functions and therefore effectively filtered out.
2. Thermal motion is treated separately and de-convoluted from the final result.
3. The resulting static density provides an effective level of comparison with theoretical results, especially if the latter has been filtered through the model by refinement of theoretical structure factors.
4. Notwithstanding the development of alternative formulations, including bond-charge models and orbital-based algorithms, no generally competitive alternative has been developed. While the pseudo atom model is widely used in experimental density analysis, it is important not to lose sight of the implied assumptions.
5. The results are dependent on the adequacy of the thermal motion formalism used (Mallinson et al. 1988), which generally is limited to the harmonic approximation.

2.7 Maximum Entropy Method

Understanding the chemical and physical properties of molecular systems requires knowledge of their charge distributions (Bader 1991). Experimentally, electron density distributions (EDD) can be reconstructed from accurate X-ray diffraction data through a series of elaborate data reduction and data analysis steps (Iversen et al. 1996). The most widely used method entails least-squares optimization of models containing atom-centered aspherical density functions (Stewart 1976; Hirshfeld 1977b; Hansen and Coppens 1978). In the empirical modeling schemes, estimates of errors of electron density and in derived properties can be calculated within the framework of the least squares method. Such estimates rely on several assumptions including the adequacy of the refined model. Several studies (Figgis et al. 1993; Chandler et al. 1994; Iversen et al. 1997) have shown that even the very sophisticated models currently used in empirical EDD modeling are inadequate to describe very fine density features present in data and in general, least-squares estimates of EDDs will therefore contain systematic bias due to the model. Nevertheless, the least-squares error estimates allow, to some extent, assessment of the reliability of conclusions drawn from the model densities.

The Maximum entropy method (MEM) is an information–theory–based technique that was first developed in the field of radio astronomy to enhance the information obtained from noisy data (Gull and Daniell 1978). The theory is based on the same equations that are the foundation of statistical thermodynamics. Both the statistical entropy and the information entropy deal with the most probable distribution. In the case of statistical thermodynamics, this is the distribution of the particles over position and momentum space, while in the case of information theory, the distribution of numerical quantities over the ensemble of pixels is considered.

The maximum entropy method (MEM) (Sakata and Sato 1990; Collins 1982; Bricogne 1988) has been introduced in charge density reconstruction. MEM can yield a high-resolution density distribution from a limited number of diffraction data. The obtained density distribution gives detailed structure information without using structural model. The ability of MEM in terms of a model-free reconstruction of the charge densities from measured X-ray diffraction data can be interpreted as "imaging of diffraction data" (Sakata and Takata 1996). From limited numbers of X-ray diffraction data, EDDs have been reconstructed by MEM in a number of systems. Maps that qualitatively reveal bonding features have been obtained. The electron-density map from MEM is one of the accurate methods for structure analysis. A precise electron-density map can be obtained and the existence of bonding electrons is clearly visible in the maximum-entropy map. The resolution of the maximum-entropy map is much higher than the map drawn by conventional Fourier transformation (Sakata and Sato 1990).

The maximum-entropy method (MEM) of calculating charge density directly from X-ray and diffractions data is a promising approach for studying the bonding state, atomic disorder, and ion conduction in detail. In X-ray diffraction, the observed crystal structure factors applied to the inverse Fourier transform are known to result in the charge density. However, this relationship between the observed crystal structure factors and charge density is premised on obtaining all observed crystal structure factors without any measurement errors. In contrast, MEM can improve these disadvantages by inferring the unobserved crystal structure factors from the observed crystal structure factors and maximizing the information entropy. MEM has been reported in various fields. These reports have provided a positive impetus for using MEM analysis to study charge density, ion conduction, and atomic disorder (Itoh et al. 2010).

In the refinement a rigid body model and restraints for intermolecular distances and appropriate angles are applied. The integrated intensities of each reflection are evaluated from the observed diffraction patterns using the result of Rietveld refinement. The final charge density is obtained after several iterative refinements.

The method is thus developed by combining the MEM with the Rietveld method to create a new sophisticated method of structure refinement in charge density level, the MEM/Rietveld method (Takata et al. 1995). The MEM/Rietveld analysis is an iterative way in combination with the MEM and Rietveld analyses. When the MEM charge density at a certain iteration step can provide a better structural model for the Rietveld of the next iteration, the iteration process continues. In this method, the final MEM electron density distribution derived is compatible with the structural model used in the Rietveld refinement (Takata et al. 2001).

2.7.1 *Maximum Entropy Enhancement of Electron Densities*

The accurate electron density distribution could be obtained if all the structure factors are known without any ambiguities. It is however impossible to collect exact values of all the structure factors by X-ray diffraction methods. The number of observed structure factors by the experiment is always limited and has some errors. The uncertainties in the results due to the incompleteness of the experimental information must be rectified. Maximum entropy method is one of the appropriate methods in which the concept of entropy is introduced to handle the uncertainties properly. The principle of MEM is to obtain electron density distribution which is consistent with the observed structure factors and to leave the uncertainties to a minimum. The mathematical formalism of MEM is given in the following lines.

The probability of a distribution of N identical particles over m boxes, each populated by n_i particles, is given by

$$P = \frac{N!}{n_1!n_2!n_3!.\,.\,.\,.\,.\,.n_m!} \tag{2.35}$$

As in statistical thermodynamics, the entropy is defined as ln P. Since the numerator is constant, the entropy is, apart from a constant, equal to

$$S = -\sum_i n_i \ln n_i \tag{2.36}$$

where Stirlings' formula (ln N! = N ln N minus; N) has been used.

In case there is a prior probability q_i for box i to contain n_i particles, then becomes

$$P = \frac{N!}{n_1!n_2!n_3!.\,.\,.\,.\,.\,.n_m!} \times q_1^{n_1} q_2^{n_2} .\,.\,.\,.\,.\,. q_m^{n_m} \tag{2.37}$$

which gives, for the entropy expression,

$$S = -\sum_i n_i \ln n_i + \sum_i n_i \ln q_i = -\sum_{i=1}^{m} n_i \ln \frac{n_i}{q_i} \tag{2.38}$$

The maximum entropy method was first introduced into crystallography by Collins (1982), who based on Eq. 2.5, expressed the information entropy of the electron density distribution as a sum over M grid points in the unit cell, using the entropy formula (Jaynes 1968).

$$S = -\sum \rho'(r) \ln \left(\frac{\rho'(r)}{\tau'(r)} \right) \tag{2.39}$$

where both $\rho'(r)$ and prior probability $\tau'(r)$ are related to the actual electron density in a unit cell as

$$\rho'(r) = \frac{\rho(r)}{\sum_r \rho(r)} \quad \text{and} \quad \tau'(r) = \frac{\tau(r)}{\sum_r \tau(r)} \tag{2.40}$$

where $\rho(r)$ and $\tau(r)$ are the electron density and prior electron density at a fixed r in a unit cell respectively. In the present theory, the actual densities are treated hereafter instead of normalized densities, and $\rho'(r)$ becomes $\tau'(r)$ when there is no information. The $\rho'(r)$ and $\tau'(r)$ are normalized as

$$\sum \rho'(r) = 1 \text{ and } \sum \tau'(r) = 1 \tag{2.41}$$

The entropy is maximized subject to the constraint

$$C = \frac{1}{N} \sum_k \frac{|F_{cal}(H) - F_{obs}(H)|^2}{\sigma^2(H)} \tag{2.42}$$

where N is the number of reflections used for MEM analysis, $\sigma(H)$, standard deviation of $F_{obs}(H)$, the observed structure factor and $F_{cal}(H)$ is the calculated structure factor given by

$$F_{cal}(H) = V \sum_r \rho(r) \exp{(-2\pi i H.r)} dV \tag{2.43}$$

where V is the volume of the unit cell.

The constraint is sometimes called a weak constraint, in which the calculated structure factors agree with the observed ones as a whole when C becomes unity. From Eq. 2.43, it can be seen that the structure factors are given by the Fourier transform of the electron density distribution in a unit cell. In the MEM analysis, there is no need to introduce the atomic factors, by which the structure factors are normally written. It should be emphasized here that it would be an assumption to use the atomic form factors in the formulation of the structure factors. Eq. 2.43 guarantees that it is possible to allow any kind of deformation of the electron densities in real space as long as information concerning such a deformation is included in the observed data.

We use Lagrange's method of undetermined multiplier (λ) in order to constrain the function C to be unity while maximizing the entropy.

We then have

$$Q = S - \frac{\lambda}{2} C \tag{2.44}$$

$$= -\sum \rho'(r) \ln\left(\frac{\rho'(r)}{\tau'(r)}\right) - \frac{\lambda}{2N} \sum_k \frac{|F_{cal}(H) - F_{obs}(H)|^2}{\sigma^2(H)} \tag{2.45}$$

And when $\frac{dQ}{d\rho} = 0$ and using the approximation,

$\ln x = x - 1$ we get,

$$\rho(r_i) = \tau(r_i) \exp\left\{\left(\frac{\lambda F_{000}}{N}\right)\left[\sum \frac{1}{\sigma(H)^2}\right]|F_{\text{obs}}(H) - F_{\text{cal}}(H)| \exp(-2\pi jH.r)\right\}$$

$$(2.46)$$

where $F_{000} = Z$, the total number of electrons in a unit cell. Eq. 2.46 cannot be solved as it is, since $F_{\text{obs}}(H)$ is defined on $\rho(r)$. In order to solve Eq. 2.46 in a simple manner, we introduce the following approximation which replaces $F_{\text{cal}}(H)$ as

$$F_{\text{cal}}(k) = V \sum \tau(r) \exp(-2\pi iH.r)\text{dV} \qquad (2.47)$$

This approximation can be called zeroth order single pixel approximation (ZSPA). By using this approximation the right-hand side of Eq. 2.46 becomes independent of $\tau(r)$ and Eq. 2.46 can be solved in an iterative way starting from a given initial density for the prior distribution. As the initial density for the prior density $\tau(r)$, a uniform density distribution is employed in this work

$$\tau(r) \leq \tau(r) \geq \frac{Z}{M} \qquad (2.48)$$

where M is the number of pixels for which the electron density is calculated. The reason for this choice of prior distribution is that uniform density distribution corresponds to the maximum entropy state among all possible density distributions. In the calculation of $\rho(r)$, all of the symmetry recruitments are satisfied and the number of electgrons (Z) is always kept constant through an iteration process. Mathematically, the summation concerning $\rho(r)$ in the above equations should be written as an integral. Since we must use a very limited number of pixels in the numerical calculation, the integral is replaced by the summation in the above equations.

After completion of the MEM enhancement, it becomes possible to evaluate the reflections missing from the summation. In a Fourier summation, the amplitudes of the unobserved reflections are assumed to be equal to zero, while the MEM technique provides the most probable values.

When extinction is present in the data set, it must be corrected before the MEM procedure is started. The structure factors must similarly be corrected for anomalous scattering, if present. Both corrections require a model for their evaluation. The independent-atom model is usually adequate for this purpose.

The advantage of MEM is a statistical deduction that can yield a high resolution density distribution from a limited number of diffraction data without using a structural model. It has been suggested that MEM would be a suitable method for examining electron densities in the inner atomic region, for example, bonding region. It gives less biased information about the electron densities as compared to conventional Fourier synthesis.

2.7.2 MEM Refinement Strategies

The technological advances in recent years bring demands for integrated 3D visualization systems to deal with both structural models and volumetric data, such as electron and nuclear densities. The crystal structures and spatial distribution of various physical quantities obtained experimentally and by computer simulations should be understood three dimensionally.

Once the structure factors are refined, they are further utilized for the evaluation of MEM charge density.

The MEM charge density calculations are done on the same formalism that Collins (1982) had adopted. In the refinement process, the analysis was performed for all data sets using Fortran 90 program PRIMA (Izumi and Dilanian 2002), to get a 3D density file. PRIMA (Practice Iterative MEM Analyses) is a MEM analysis program to calculate electron densities from X-ray diffraction data. The input file contains the cell parameters, space group, pixels, total charge, Lagrange parameter, and structure factors. In the present work, the unit cell was divided into 64^3 pixels and the initial electron density at each pixel was fixed uniformly as Z/a_0^3, where Z is the number of electrons in the unit cell. The electron density is evaluated by carefully selecting the Lagrange multiplier in each case such that the convergence criterion C becomes unity after performing minimum number of iterations. The 3D electron density was plotted using VESTA (Visualization of Electron/Nuclear densities and Structures) (Momma and Izumi 2006) software package. VESTA is a 3D visualization program for structural models and 3D grid data such as electron/nuclear densities. VESTA deals with structural models and volumetric data at the same window. Virtually unlimited number of objects such as atoms, bonds polyhedra, and polygons on iso-surfaces are dealt with. Visualization of inter atomic distances and bond angles are possible. Transparent iso-surfaces can be overlapped with structural models.

2.8 Pair Distribution Function

The method of crystallographic analysis of atomic structure is so well established today that we often fail to see the enormity of the feat accomplished by a simple X-ray diffraction measurement, the positions of as many as 10^{23} atoms are determined with an accuracy of 10^{-4} nm. This is the benefit of lattice periodicity. But modern materials are often disordered and standard crystallographic methods lose the aperiodic (disorder) information. The structural analysis of non-periodic matter is more difficult than crystalline solids. Recent advances in the experimental methods, have greatly improved our ability to characterize the atomic structure of non-periodic matter using the atomic pair distribution function (PDF) analysis and to find local structural deviations from a well-defined average structure (Egami 1990; Billinge et al. 1996).

By definition, the atomic pair distribution function PDF is the instantaneous atomic density–density correlation function which describes the atomic arrangement in materials (Billinge et al. 2000). It is the sine Fourier transform of the experimentally observable structure factor obtained in a powder diffraction experiment (Warren 1990). Since the total structure function includes both the Bragg intensities and diffuse scattering its Fourier associate, the PDF, yields both the local and average atomic structure of materials. By contrast, an analysis of the Bragg scattering intensities alone yields only the average crystal structure. Determining the PDF has been the approach of choice for characterizing glasses, liquids and amorphous solids for a long time (Wagner 1978). However, its widespread application to crystalline materials, like manganites, where some local deviation from the average structure is expected to take place, has been relatively recent (Egami 1998).

We chose to use high-energy X-rays to measure the PDFs because it is possible to get high-quality data at high-Q values (Q is the magnitude of the wave vector) allowing accurate high real-space resolution PDFs to be determined (Petkov et al. 1999a, b). It was previously thought that neutrons were superior for high-Q measurements because, as a result of the Q dependence of the X-ray atomic form factor the X-ray coherent intensity gets rather weak at high Q; however, the high flux of X-rays from modern synchrotron sources more than compensates for this and we have shown that high quality high-resolution PDFs can be obtained using X-rays.

The study of alloys is complicated by the fact that considerable local atomic strains are present due to the disordering effect of the alloying. This means that local bond lengths can differ from those inferred from the average crystallographic structure by as much as 0.1 Å. This clearly has a significant effect on calculations of electronic and transport properties. To characterize the structure of alloys fully, it is necessary to augment crystallography with local structural measurements. In the past the extended X-ray absorption fine structure (XAFS) technique has been extensively used. More recently, the atomic pair distribution function (PDF) analysis of powder diffraction data has also been applied to obtain additional local structural information (Peterson et al. 2001).

2.8.1 Atomic Pair Distribution Function

The elastic and quasi-elastic scattering intensity of X-rays or neutrons by a collection of atoms, after correction for absorption, polarization and multiple scattering and normalizing to the unit of one atom or one electron scattering (Warren 1990) is given by

$$I(Q) = \sum_{ij} f_i(Q) f_j(Q) \langle \exp\left(iQ.\left(r_i - r_j\right)\right) \rangle \qquad (2.49)$$

where $f_i(Q)$ is the scattering amplitude of a single atom i, r_i is the position of the ith atom, $\langle \exp\left(iQ.(r_i - r_j)\right)\rangle$ is the quantum and thermal average, and Q is the scattering vector,

$$Q = k_f - k_i \tag{2.50}$$

$$Q = |Q| = 2|k_{fi}| \sin\theta \quad (if\, |k_f| = k_i) \tag{2.51}$$

where k_f and k_i are the momenta of the scattered and incident photons or neutrons, respectively, and θ is the diffraction angle. For X-ray scattering the quasi-elastic scattering intensity usually includes the inelastic scattering intensity due to phonons, since unless a special high resolution setup is utilized it cannot be separated from the elastic scattering, while the single-electron inelastic (Compton) scattering intensity is excluded. For neutron scattering the phonon is either excluded or included only approximately. The average, or total, structure factor can be defined as

$$S(Q) = \frac{I(Q)}{\langle f(Q)\rangle^2} + \frac{\left[\langle f(Q)\rangle^2 - \langle f(Q^2)\rangle\right]}{\langle f(Q)\rangle^2} \tag{2.52}$$

where $\langle f(Q)\rangle^2$ is the compositional average of structure factor.

If $S(Q)$ is isotropic, i.e., independent of the orientation of Q, the Fourier transformation of (2) in three dimensions gives the atomic pair distribution function (PDF),

$$\tilde{n}(r) = \tilde{n}_0 + \frac{1}{2\eth^2 r} \int Q[S(Q) - 1] \sin(Qr) dQ, \tag{2.53}$$

where $\tilde{n}_0$ is the average atomic number density. When the unit of Q is nm^{-1}, the unit of $\tilde{n}(r)$ is nm^{-3}, and $\tilde{n}(r)$ corresponds to the atomic number density at a distance r from an average atom.

$$4\eth r(\tilde{n}(r) - \tilde{n}_0) = \frac{2}{\eth} \int Q[S(Q) - 1] \sin(Qr) dQ \tag{2.54}$$

The frequently used atomic PDF, also called $G(r)$ is defined as,

$$G(r) = 4\eth r(\tilde{n}(r) - \tilde{n}_0) \tag{2.55}$$

where $G(r)$ gives the number of atoms in a spherical shell of unit thickness at a distance r from a reference atom. It peaks at characteristic distances separating pairs of atoms and thus reflects the atomic structure. $G(r)$ is the Fourier transform of the experimentally observable total structure function, $S(Q)$ that is

$$G(r) = \frac{2}{\eth} \int_{Q=0}^{Q=\max} Q[S(Q) - 1] \sin(Qr) dQ \tag{2.56}$$

where Q is the magnitude of the wave vector.

The structure function is related to the coherent part of the total diffracted intensity of the

$$S\left(\vec{Q}\right) = 1 + \frac{I^{\mathrm{coh}}\left(\vec{Q}\right) - \sum c_i \left| f_i\left(\vec{Q}\right)\right|^2}{\left|\sum c_i f_i\left(\vec{Q}\right)\right|^2} \tag{2.57}$$

where $I^{\mathrm{coh}}(Q)$ is the coherent scattering intensity per atom in electron units and c_i and f_i are the atomic concentration and X-ray scattering factor, respectively, for the atomic species of type i. $G(r)$ is simply another representation of the diffraction data. However exploring the diffraction data in real space is advantageous, especially in the case of materials with significant structural disorder.

In order to refine an experimental PDF one needs to calculate a PDF from a structural model. This can be done using the relation

$$G_{\mathrm{calc}}(r) = \frac{1}{r} \sum_i \sum_j \left[\frac{b_i b_j}{\langle b \rangle^2} \ddot{a}(r - r_{ij}) \right] - 4 \delta r \tilde{n}_0 \tag{2.58}$$

where the sum goes over all pairs of atoms i and j within the model separated by r_{ij}. The scattering power of atom i is b_i and $\langle b \rangle$ is the average scattering power of the sample. In case of neutron scattering, b_i is simply the scattering length, in case of X-rays, it is the atomic form factor evaluated at a user defined value of Q.

2.8.2 Important Details of the PDF Technique

1. The total scattering, including Bragg scattering as well as diffuse scattering, contributes to the PDF. In this way both the long-range atomic structure, manifested in the sharp Bragg peaks, and the local structural imperfections, manifested in the diffuse components of the diffraction pattern, are reflected in the PDF.
2. $G(r)$ is barely influenced by diffraction optics and experimental factors since these are accounted for in the step of extracting the coherent intensities from the raw diffraction data. This renders the PDF a structure-dependent quantity only.
3. By accessing high values of Q, experimental $G(r)$ values with high real-space resolution can be obtained and, hence, quite fine structural features are revealed (Petkov et al. 1999a, b). In fact data at high Q values ($Q > 15$) are critical to the success of PDF analysis. Therefore, the PDF can serve as a basis for structure determination.

Once the PDF is obtained an approach similar to Rietveld refinement is followed. A model atomic configuration is constructed and the respective PDF calculated and compared with the experimental one. Structural parameters in the model such as atomic positions, thermal factors, and occupancies are then varied in

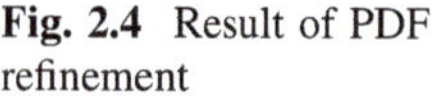

Fig. 2.4 Result of PDF refinement

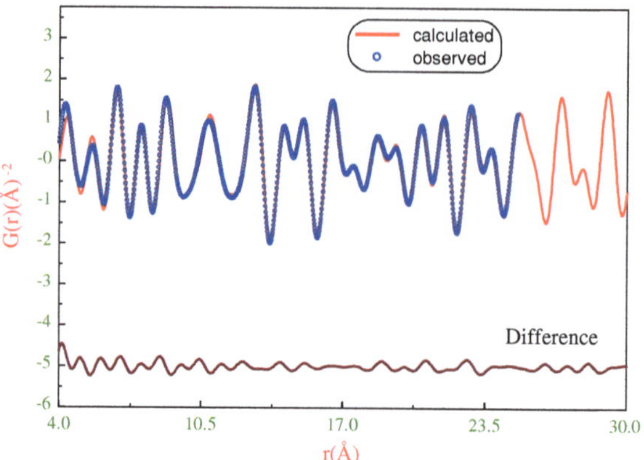

such a way as to improve the agreement between the calculated and experimental PDFs. This is done with, or without, observing predefined constraints imposed by the symmetry of the space group of the crystal structure being tested (Proffen and Billinge 1999). In this way, local distortions away from the average structure or lower, unresolved, symmetries can be modeled. The PDF is very sensitive to the coordination environment of atoms over short (<5 Å) and intermediate (5–20 Å) ranges. The approach has proved to be quite successful in determining the structure of various crystalline materials exhibiting different degrees of structural disorder (Gutmann et al. 2000; Proffen et al. 1999; Petkov et al. 1999a, b).

In Fig. 2.4 the result of PDF refinement is plotted against $G(r)$ (the number of atoms in a spherical shell of unit thickness) and r (the distance from a reference atom). The solid line is the calculated PDF. The fitted circles are the observed data. The difference between the calculated and observed PDF is plotted below the data.

Since PDF contains Bragg and diffuse scattering, the information about local arrangements is preserved. The PDF can be understood as a bond—length distribution between all pairs of atoms i and j within the crystal (up to a maximum distance), however each contribution has a weight corresponding to the scattering power of the two atoms involved.

2.8.3 Calculation of PDF

The study of a measured PDF ranges from a simple peak width analysis revealing information about correlated motion (Jeong et al. 1998) to the full profile refinement of the PDF based on a structural model either using the reverse Monte Carlo technique (Toby and Egami 1992) or least-squares regression (Billinge 1998) as implemented in the program PDFFIT (Proffen and Billinge 1999).

The experimental PDF was obtained by initially extracting the coherently scattered intensities from X-ray diffraction pattern by applying appropriated correction for flux, Compton scattering, and sample absorption. The intensity was normalized in absolute electron unit, reduced to atomic PDF. All data procession was done using the program PDFgetX (Jeong et al. 2001).

PDFgetX (Jeong et al. 2001) is a program used to obtain the observed atomic pair distribution function (PDF) from measured X-ray powder diffraction data. The observed and calculated PDF has been obtained from the graphical software PDFgui (Farrow et al. 2007), which is a graphical environment for PDF fitting. This allows for powerful usability features such as real-time plotting and remote execution of the fitting program while visualizing the results locally. Modeling of PDF was done using the software PDFFIT (Proffen and Billinge 1999) to yield structural parameters.

The program PDFFIT is designed for the full profile structural refinement of the atomic pair distribution function (PDF). In contrast to conventional structure refinement based on Bragg intensities, the PDF probes the local structure of the studied material. The program allows the refinement of atomic positions, aniso-tropic thermal parameters, and site occupancies as well as lattice parameters and experimental factors. By selecting individual atom types one can calculate partial and differential PDFs in addition to the total PDF. Furthermore, one can refine multiple data sets and/or multiple structural phases. The program is controlled by a command language, which includes a FORTRAN style interpreter supporting loops and conditional statements. This command language is also used to define the relation between refinement parameters and structural or experimental infor-mation, allowing virtually any constraint to be implemented in the model. PDFFIT is written in Fortran-77. The basic concept, command language, and some file formats of PDFFIT are taken from the diffuse scattering and defect structure simulation program DISCUS (Proffen and Neder 1997).

Using the PDFFIT (Proffen and Billinge 1999) software the PDF refinement was executed with the refinement of structural parameters like lattice parameters, phase scale factor, linear atomic correlation factor, quadratic atomic correlation factor, low r peak sharpening, peak sharpening cut-off and cut-off, for profile setup functioning to get the absolute phase. The data configuration parameters are PDFfit range with step size, data scale factor, upper limit for Fourier transform to obtain data PDF, resolution peak broadening factor, data collection temperature etc., which can be refined to get accurate PDF fitting. In the end the observed and calculated PDFs are visualized and compared.

2.8.4 Significance of PDF

- We would like to be able to sit on an atom and look at our neighborhood. The PDF method allows us to do that.
- The PDF gives different information about different length-scales. We can see the structure within a domain at low-r and between domains at high-r.
- PDF gives both local and average structure information.

References

Azaroff LV (1968) Elements of x-ray crystallography. Mc Graw hill book company, New York, p 79

Bader RFW (1991) Atoms in molecules, a quantum theory. Oxford University Press, Oxford

Berg JE, Werner PE (1977) Z. Kristallogr 145:310–320

Billinge SJL (1998) Local structure from diffraction. In: Billinge SJL, Thorpe MF. Plenum, New York, p 137

Billinge SJL, DiFrancesco RG, Kwei GH, Neumeier JJ, Thompson JD (1996) Phys Rev Lett 77:715–718

Billinge SJL, Proffen TH, Petkov V (2000) Phys rev B 1 62(Number 2):1203–1211

Bragg WL (1912) Nature 90:410–410

Bragg WL (1913) The diffraction of short electromagnetic waves by a crystal. Proc Cambridge Philoso Soc 17:43–57

Breck DW, Everslole WG, Milton RM, Reed TB, Thomas TL (1956) J Am Chem Soc 78:5963–71

Bricogne G (1988) Acta Cryst A44:517–545

Caglioti G, Paeletti A, Ricci FP (1958) Nucl Instrum 3:223–228

Chandler GS, Figgis BN, Reynolds PA, Wolff SK (1994) Chem Phys Lett 225:421–426

Cheetham AK, Taylor JC (1977) J Solid State Chem 21:253–257

Chen R, Trucano P, Stewart RF (1977) Acta Cryst A33:823–828

Collins DM (1982) Nature 298:49–51

Coppens P, Yang YW, Blessing RH, Cooper WF, Larsen FK (1977) J Am Chem Soc 99:760–766

Coppens P, GuruRow TN, Leung P, Stevens, ED, Becker PJ, Yang YW (1979) Acta Cryst A 35:63–72

Cotton FA, Stanley TT (1977) Inorg Chem 16:2668–2671

Cox DE, Hastings JB, Thomlinson W, Prewill CT (1983) Nucl Instrum Methods 208:573–578

Cromer DT, Larsen AC, Stewart RF (1976) J Chem Phys 65:336–349

Cullity BD, Stock SR (2001) Elements of X-ray diffraction, Pearson education. 3rd edn. Prentice Hall, Upper Saddle River, p 558

Dawson B (1967) Proc Royal Soc London Ser A 298:255–263

De Marco JJ, Weiss RJ (1965) Phys Rev 137 A1:869–871

Dinnibier RE (2008) Powder diffraction: theory and practice. In: Dinnibier RE, Billinge SJL (eds.) Royal society of chemistry, p 5

Dollase WAJ (1986) Appl Crystallogr 19:267–272

Egami T (1990) Mater Trans JIM 31:163–176

Egami T, Billinge SJL, Thorpe MF (eds) (1998) Local structure from diffraction. Plenum, New York

Farrow CL, Juhas P, Liu JW, Bryndin D, Bozin ES, Bloch J, Proffen T, Billinge SJL (2007) J Phys Condens Matter 19:335219

Figgis BN, Iverson BB, Larsen FK, Reynolds PA (1993) Acta Cryst B49:794–806

Fox AG, Tabernor MA (1991) Acta Metall Mater 39:669–678

George W (2006) Powder diffraction—The Rietveld method and two stage method chapter 1. Springer, Berlin

Goeta AE, Howard JAK (2004) Chem Soc Rev 33:490–500

Gull SF, Daniell GJ (1978) Nature 272:686–690

Guru Row TN (1983) J Chem Sci 92:4–5

Gutmann M, Billinge SJL, Brosha EL, Kwei GH (2000) Phys Rev B 61:11762–11768

Hansen NK, Coppens P (1978) Acta Cryst A34:909–921

Hansen NK, Schneider JR, Yellon WB (1987) Pearson WH Acta Cryst A43:763–769

Harel M, Hirshfeld FL (1975) Acta Cryst B31:162–172

Hewat AW (1986) Chem Scripta 26A:119–130

Hirshfeld FL (1971) Acta Cryst B 27:769–781

Hirshfeld FL (1977a) Isr J Chem 16:226–229

Hirshfeld FL (1977b) J Chem 16:226–229

Howard CJ (1982) J Appl Crystallogr 15:615–620

Itoh T, Shirasaki S, Fujie Y, Kitamura N, Idemoto Y, Osaka K, Ofuchi H, Hirayama S, Honma T, Hirosawa I (2010) J Alloys Compd 491:527–535

Iversen BB, Larsen FK, Figgis BN, Reynolds PA (1996) Acta Cryst B53:923–932

Iversen BB, LarsenFK, Figgis BN, Reynolds PA (1997) J Chem Soc 2227–2240

Izumi F, Dilanian RA (2002) Recent research developments in physics Part II, vol vol 3. Transworld Research Network, Trivandrum, pp 699–726

Jaynes ET (1968) IEEE Trans Syst Sci Cybern SSC-4:227–241

Jeong IK, Proffen Th, Jacobs F, Billinge SJL (1998) J Phys Chem A 103:921–924

Jeong IK, Thompson J, Proffen T, Perez A, Billinge SJL (2001) PDFGetX a program for obtaining the atomic pair distribution function from X-ray powder diffraction data. J Appl Cryst 34:536

Kokatalio GT, Lawton SL, Olson DH, Meier WM (1978) Nature 272:437–438

Kurki-Suonio K (1968) Acta Cryst A 24:379–390

Larsen FK, Lehmann MS, Merisalo M (1980) Acta Cryst B40:159–163

Le Bail A, Duroy H, Fourquet JL (1988) Mater Res Bull 23:447–452

Mallinson PR, Koritsanszky T, Elkaim E, Li N, Coppens P (1988) Acta Cryst A 44:336–342

Malmros G, Thomas JO (1977) J Appl Crystallogr 10:7–11

March A (1932) Z Kristallogr 81:285–297

Marin C, Dieguez E (1999) Orientation of single crystals by back reflection laue pattern simulation. World Scientific Publishing, Singapore, p 23

Mcconnell JF, Sanger PL (1970) Acta Cryst A 26:83–93

Momma K, Izumi F (2006) Comm Crystallogr Comput IUCr Newslett 7:106–119

Novelline RA, Squire LF (2004) Squire's fundamentals of radiology. chapter 1, Harvard University press, Cambridge

Pawley GS (1981) J Appl Crystallogr 14:357–361

Peterson PF, Proffen T, Jeong IK, Billinge SJL, Choi KS, Kanatzidis MG, Radaelli PG (2001) Phys Rev B 63:165211–165218

Petkov V, Jeong IK, Chung JS, Thorpe MF, Kycia S, Billinge SJL (1999a) Phys Rev Lett 83:4089–4092

Petkov V, Francesko DRG, Billinge SJL, Acharaya M, Foley HC (1999b) Phil Mag B79:1519–1530

Petříček V, Dušek M, Palatinus L (2000) JANA 2000. The crystallographic computing system Institute of Physics Academy of sciences of the Czech republic, Praha

Petříček V, Dušek M, Palatinus L, JANA (2006) The crystallographic computing system. Institute of Physics Academy of sciences of the Czech republic, Praha

Price PF, Malsen EN (1978) Acta Cryst A34:173–183

Proffen T, Neder R (1997) DISCUS a program for diffuse scattering and defect structure simulations. J Appl Cryst 30:171–175

Proffen Th, Billinge SJL (1999) J Appl Cryst 32:572–575

Proffen Th, Francesco DRG, Billinge SJL, Brosha EL, Kwei GH (1999) Phys Rev B60:9973–9977

Purrington RD (1997) Physics in the nineteenth century, history of science/physics. Rutgers university press, New Brunswick, p 164

Rees B (1976) Acta Cryst A32:483–488

Rees B (1977) Isr J Chem 16:180–186

Rietveld HM (1969) J Appl Crystallogr 2:65–71

Ruben AD, Izumi F (2004) Super-fast program PRIMA for the maximum-entropy method, advanced materials laboratory, National institute for materials science. 1-1 Namiki, Tsukuba, Ibaraki, Japan 305:0044

Sakata M, Sato M (1990) Acta Cryst A46:263–270

Sakata M, Takata M (1996) High Press Res 14:327–333

Shmueli U (2008) IUCr series, International tables of crystallography. Volume 2, chapter1, 3rd edn. Springer, Berlin

Stevens ED, Coppens P (1976) Acta Cryst A32:915–917

Stewart RF (1969) J Chem Phys 51:4569–4577

Stewart RF (1972) J Chem Phys 57:1664–1668

Stewart RF (1976) Acta Cryst A32:545–574

Stout GH, Jensen LH (1968) X-ray structure determination—a practical guide. The Macmillan Company Collier-Macmillan, London, p 217

Takata M, Umeda B, Nishibori E, Sakata M, Saito Y, Ohno M, Shinohara H (1995) Nature 377:46–49

Takata M, Nishibori E, Shinmura M, Tanaka H, Tanigaki K, Kosaka M, Sakata M (2001) Mater Sci Eng A312:66–71

Thompson P, Cox DE, Hastings JB (1987) J Appl Crystallogr 20:79–83

Toby BH, Egami T (1992) Acta Cryst A48:336–346

Wagner CNJ (1978) J Non Cryst Solids 31:1–40

Warren BE (1990) X-ray diffraction, chapter 3. Dover publications, New York

Weiss H (1925) Proceedings of the royal society of london series A 108(748):643–654

Wertheim GK, Butler MA, West KW, Buchanan DNE (1974) Rev Sci Instrum 45:1369–1371

Young RA (1995) The rietveld method, Volume 5 of International union of crystallography momographs on crystallography, Chap 1. Oxford Science publications, Reprint Oxford University press

Young RA, Mackie PE, Von Dreele RBJ (1977) Appl Crystallogr 10:262–269

Zachiarsen WH (1948) Acta Crstallogr 1:265–8, 277–87

Zachiarsen WH, Ellinger FH (1963) Acta Crystallogr 16:369–75

Chapter 3
Results and Discussion on Metals and Alloys

Abstract Through centuries of scientific investigation we have come to understand that all materials are composed of atoms and electrons. The number of electrons surrounding the nucleus defines the characteristics of that atom. Atoms are held together through electrons in forming a material. Our experience convinces that there are two types of electron bonding (a) strong bonding and (b) weak bonding. Strong bonding involves direct electron–electron bonding and can be differentiated as covalent, ionic and metallic bonding. Weak bonding utilises Van der Waals' forces or dipole–dipole interaction in forming bonding. The understanding of the bonding may be enhanced by visualisation of the electron bonding system. This chapter deals with effective analysis of elemental metals, alloys and dilute doped alloys through versatile techniques such as the maximum entropy method (Collins 1982), multipole (Hansen and Coppens 1978) and atomic pair distribution function (Proffen and Billinge 1999). This book deals with the bonding and electron density distribution of the following metals and alloys through X-ray single crystal and powder analysis. 1. Sodium and vanadium, 2. Aluminium, nickel and copper, 3. Magnesium, titanium, iron, zinc, tin and tellurium, 4. Cobalt aluminium and nickel aluminium, 5. Nickel chromium ($Ni_{80}Cr_{20}$), 6. $Na_{1-x}Ag_xCl$ (x = 0.03 and 0.10), 7. Aluminium, with iron impurities (0.215 and 0.304 wt% Fe).

3.1 Sodium and Vanadium Metals

3.1.1 Introduction

The study of bonding in metals is interesting because of the various of uses of metals. The metals sodium and vanadium have found extensive use in the fields of chemistry, industry, research etc. Sodium and vanadium are useful in their elemental and compound forms.

R. Saravanan and M. Prema Rani, *Metal and Alloy Bonding: An Experimental Analysis,*   65
DOI: 10.1007/978-1-4471-2204-3_3, © Springer-Verlag London Limited 2012

The precise study of bonding in materials is always useful and interesting yet no study can reveal the real picture, because no two experimental data are identical. This problem is enhanced when the model used for the evaluation of electron densities is not entirely suitable. Fourier synthesis of electron densities can be of use in picturing the bonding between two atoms, but it suffers from the major disadvantage of series termination error and negative electron densities which prevent the clear understanding of the bonding, a factor intended for analysis. The advent of Maximum Entropy Method (MEM) has solved many of these problems. MEM electron densities are always positive and even with limited data, one can determine reliable electron densities resembling the true densities. Currently, multipole analysis of charge densities has been widely used to study crystalline materials (Coppens and Volkov 2004; Poulsen et al. 2004; Frils et al. 2004; Pillet et al. 2004; Marabello et al. 2004). The multipole technique of synthesising the electron density of an atom into core and valence parts yields an accurate picture of the bonding in a crystalline system. In this work, such an analysis has been carried out using the formalism of Hansen and Coppens (1978).

Bonding in materials has been analysed by Israel et al. (2002), Saravanan et al. (2002a, b), Kajitani et al. (2001) using MEM technique and reported X-ray data and they obtained precise information about bonding in materials. The nature of bonding in these materials as analysed are found to be ionic, mixed covalent and ionic, and 'oxide' bonding.

3.1.2 Summary of the Work

The nature of bonding and the charge distribution in sodium and vanadium metals have been analysed using the reported X-ray data of these metals. MEM and multipole analysis have been used for elucidation and analysis of bonding in these metals. The mid-bond densities in sodium and vanadium are found to be 0.014 and 0.723 e/Å^3 respectively, giving an indication of the strength of the bonds in these materials. From the multipole analysis, the sodium atom is found to contract more than the vanadium atom.

3.1.3 Origin of the Data

In this study on sodium and vanadium, we have attempted to study the metal bonding in these metals using MEM and multipole techniques. The X-ray structure factor data of sodium has been taken from Field and Medlin (1974) and that of vanadium from Linkoaho (1972). Other relevant details can be found from these papers. Sodium has been chosen in this present study because of its very large thermal vibration (Field and Medlin 1974) and because it would be interesting to analyse the highly reactive nature of the metal in terms of strength of bonding and thermal smearing of valence region. In order to make a comparison of the results with an element having the same structure, vanadium has been chosen, which has a very low thermal vibration parameter (Linkoaho 1972). Other factors such as the quality of the data have also been taken into account in selecting these systems. Hansen and Coppens (1978) proposed a modified electron density model with the option that allows the refinement of population parameters at various orbital levels where the atomic density is described as a series expansion in real spherical harmonic functions through fourth order Y_{lm}. According to this model, the charge density in a crystal is written as the superposition of harmonically vibrating aspherical atomic density distribution convolving with the Gaussian thermal displacement distribution as

$$\rho(r) = \sum_{k}^{\text{atoms}} \rho_k\left(\overrightarrow{r} - \overrightarrow{r}_k - \overrightarrow{u}\right) \otimes t_k\left(\overrightarrow{u}\right), \tag{3.1}$$

where $t_k(\overrightarrow{u})$ is the Gaussian term and $\otimes$ indicates a convolution. The atomic charge density is then defined as

$$\rho\left(\overrightarrow{r}\right) = P_c\rho_{\text{core}}\left(\overrightarrow{r}\right) + P_v k'^3 \rho_{\text{valence}}(k'r) + \sum_{l=0}^{4} k''^3 R_l(k''r) \sum_{m=-l}^{l} P_{lm} Y_{lm}\left(\frac{\overrightarrow{r}}{r}\right) \tag{3.2}$$

where P_c, P_v and P_{lm} are population coefficients. Canonical Hartree–Fock atomic orbitals of the free atoms normalised to one electron can be used for the construction ρ_{core} and ρ_{valence} but the valence function is allowed to expand and contract by the adjustment of the variable parameters k' and k''. The effect of the temperature can be distinguished from the convoluted and the deconvoluted form of thermal contribution to the charge density as dynamic and static multipole deformation maps. The deformation density in these maps are characterised by

$$\Delta\rho_{\text{multipole\_deformation}}\left(\overrightarrow{r}\right) = \frac{1}{v}\sum\left[F\left(\overrightarrow{h}\right)_{\text{multipole}} - F\left(\overrightarrow{h}\right)_{\text{spherical\_atom}}\right]\exp\left[-2\left(\overrightarrow{h}, \overrightarrow{r}\right)\right] \tag{3.3}$$

where, $F_{\text{multipole}}$ is the Fourier transform of the multipole charge density with or without the convolution of thermal contribution, where the Fourier components are terminated at the experiment resolution.

3.1.4 Data Analysis

Both the reported data sets have been analysed using the GSAS (Larson and Von Dreele 2004) programme of the Los Alamos National Laboratory, USA. The reliability index wR for sodium data is found to be 2.2% after successful

Table 3.1 Parameters from MEM refinement

Parameters	Na	V
Prior electron density (e/Å^3)	0.2785	1.6634
Resolution (e/Å)	0.067	0.047
λ	0.024	0.024
C	1.0	1.0
R(%)	2.157	0.392
wR(%)	2.356	0.498
Number of cycles	1464	236

λ Lagrangean parameter; C Convergence criterion

refinement. The refined Debye–Waller factor is found to be $B_{Na} = 7.899$ (96) Å^2. This value is very large and comparable with that reported [$B_{Na} = 7.86$ Å^2, Field and Medlin (1974)], the difference being only 0.039 Å^2. The reliability index for the vanadium refinement is wR $= 1.1\%$ and the Debye–Waller factor is found to be 0.404(24) Å^2. This value is also comparable to that reported by Linkoaho (1972), i.e., $B_V = 0.37$ Å^2, the difference being 0.034 Å^2.

The reliability indices and the matching of thermal parameters with the corresponding reported values for both the atoms indicate that the quality of the data for the electron density analysis is suitable enough. The same number of data for both the crystal systems has been used for the present analysis in order to make a comparison also of the bonding (only insignificant changes in the density maps and numerical results has been observed on including all the reflections for sodium).

The MEM analysis has been carried out by dividing the cell into 64^3 pixels. In the case of sodium data, convergence condition C $= 1.0$ was obtained after 1464 cycles of refinement; the reliability indices are found to be $R_{MEM} = 2.157\%$ and wR$_{MEM} = 2.356\%$. The Lagrange parameter λ used was 0.024 for sodium. For vanadium, convergence was obtained after 236 cycles with R values as $R_{MEM} = 0.392\%$ and wR$_{MEM} = 0.498\%$. A value of 0.024 was used for λ. The prior densities assumed for sodium and vanadium were 0.2785 and 1.6634 e/Å^3, respectively. The resolution of the map was 0.067 and 0.047 Å/pixel for sodium and vanadium, respectively. The analysis on the MEM results has been given in the following section. Some important results from the MEM refinements are given in Table 3.1.

3.1.5 Results and Discussion

The MEM electron density maps in the density rich regions of sodium and vanadium on the (100) plane have been given in Fig. 3.1a, b, respectively.

The density distribution of sodium and vanadium on the (110) plane has been shown in Fig. 3.2a, b, respectively.

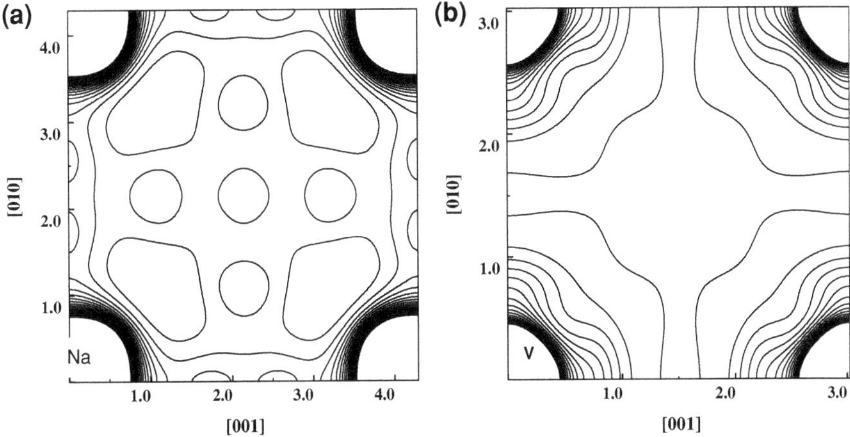

Fig. 3.1 **a** MEM electron density of sodium on (100) plane. *Contour lines* are between 0.0 and 1.0 e/Å^3. Contour interval is 0.05 e/Å^3. **b** MEM electron density of vanadium on (100) plane. *Contour lines* are between 0.2 and 4.5 e/Å^3. Contour interval is 0.224 e/Å^3

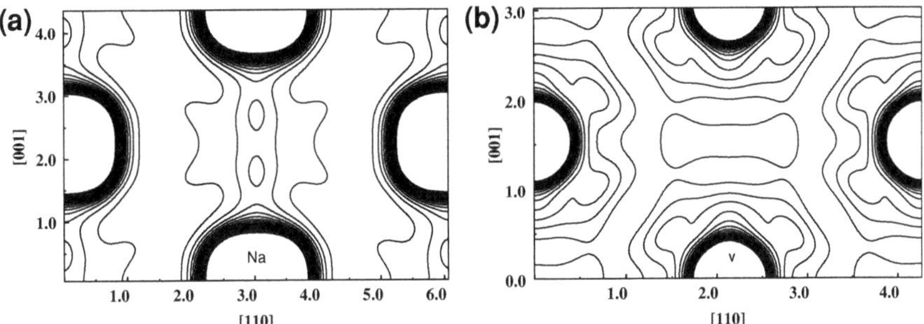

Fig. 3.2 **a** MEM electron density of sodium on (110) plane. *Contour lines* are between 0.0 and 1.0 e/Å^3. Contour interval is 0.05 e/Å^3. The plane has been shifted half a unit cell along the z direction. **b** MEM electron density of vanadium on (110) plane. *Contour lines* are between 0.04 and 6.0 e/Å^3. Contour interval is 0.298 e/Å^3. The plane has been shifted half a unit cell along the z direction

Figure 3.3a shows the one-dimensional variation of the electron densities of sodium along the three directions, [100], [110] and [111]. Figure 3.3b shows the one-dimensional variation of electron density of vanadium along the three crystallographic directions.

The Fig. 3.4a, b represent the error analysis of the data used for the present analysis.

The density maps of sodium and vanadium show highly resolved electron density contours suitable for precise analysis. Both the systems belong to the body-centred cubic structure. Hence, in sodium, only the valence electronic contribution

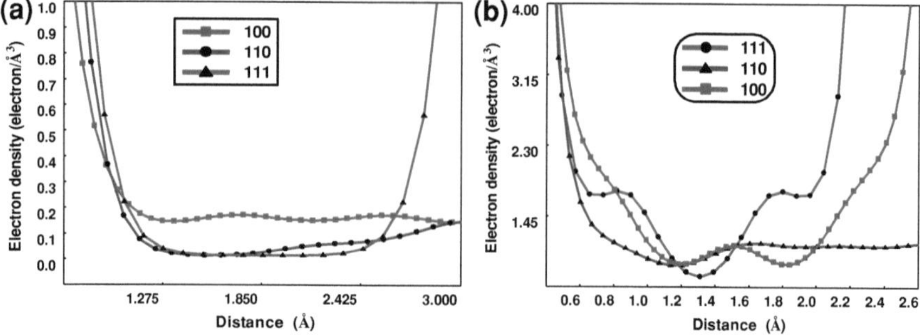

Fig. 3.3 a One-dimensional electron density variation of sodium along [100], [110] and [111] directions. **b** One-dimensional electron density variation of vanadium along [100], [110] and [111] directions

Fig. 3.4 a The histogram of number of reflections plotted in the error space of sodium. **b** The histogram of number of reflections plotted in the error space of vanadium

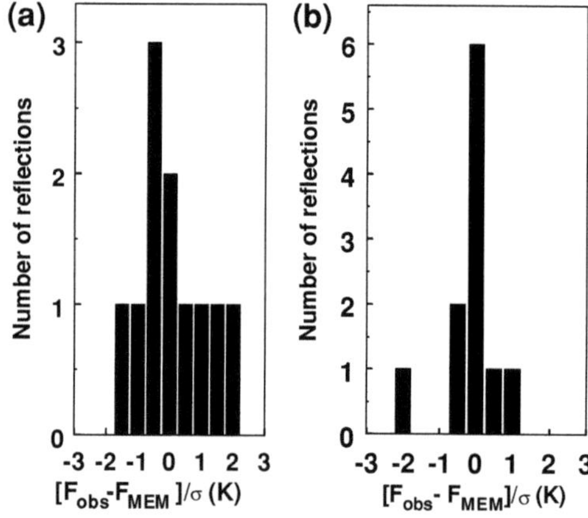

of the body-centred atom is seen at the centre of the (100) plane of the cell as seen from Fig. 3.1a. It also represents the very large thermal vibration of the sodium atom.

The corners show that the core regions are perfectly symmetric and spherical. The four voids seen on the (100) plane conform to the crystal system. The conical-shaped contour lines seem to be attracted towards the centre from the corners. This may be due to the fact that the core of the centre atom attracts the valence electrons of the eight corner atoms. Since the atomic number of sodium is only 11, this core–valence attraction is very flexible and leads to large smearing of the electrons. Moreover, it should be remembered that the thermal vibration of the sodium atom being very large ($B_{Na} = 7.899(96)$ Å^2), makes this electrostatic interaction process highly feasible. The large thermal parameter of sodium and also the smaller atomic number of sodium makes the body-centred sodium atom possible to reveal itself at the face centre.

Figure 3.1b shows the density distribution of vanadium on the (100) plane. The centre of the outer contour lines are bent towards the core of each atom, showing strong attraction of valence electrons by the core, which is due to the large atomic number of vanadium and its smaller thermal vibration parameter. Two humps are seen at the outer contour lines of each of the corner atoms. This may be due to the overlapping of charges between the cores in the bonding direction.

Figure 3.2a, b show the densities of sodium and vanadium respectively, on the (110) plane (shifted by half the cell distance along z direction). Large thermal smearing of sodium atom is seen (Fig. 3.2a) as seen from the body-centred atom. The core of one atom attracting the valence of other atom is also seen. In vanadium also, this process is taking place as seen from Fig. 3.2b.

From Fig. 3.3a, which shows the one-dimensional variation of the electron densities of sodium along the three directions, [100], [110] and [111], the minimum density is found at a distance of 1.732 Å at which the electron density is found to be 0.014 e/Å^3. The electron density along the bonding region of the [111] direction is flat with low level of density similar to the density along the [110] direction. But along [100] direction, the density is slightly higher than that along the other two directions. The large thermal vibration of sodium atom and the smallest inter atomic distance along [100] direction leads to higher density value along [100]. The inter atomic distances are larger along [110] and [111] directions and the lone valence electron of sodium atom leads to lower densities along these directions. As far as the [100] direction is concerned, the amount of electrons contributing to the bond electron density comes from the corner atom along [100] direction and the body-centred atom too (due to the large thermal smearing), to decide the electron density along the shortest [100] direction.

Figure 3.3b shows the one-dimensional variation of electron density of vanadium along the three crystallographic directions. There is a slight increasing undulation of electron density along [111] direction. This starts beyond a distance of about 0.7 Å from the centre. The electron density at 1.309 Å is found to be 0.723 e/Å^3. This is an indication of the bond strength in vanadium compared to that of sodium, which has a density of 0.014 e/Å^3. The small ellipses of charge density in the [110] plane as seen from Fig. 3.8b and the void in Fig. 3.8a confirm these magnitudes. The reported (Wilkinson) radii for sodium and vanadium are 1.57 and 1.22 Å. In this work, the radii are estimated from the electron densities along the [110] direction, because, along [100] direction, the atoms are too close to decide the inter atomic interactions and along [111] direction, the bonding effects are predominant. Thus, the estimated radii are 1.517 and 1.136 Å respectively, for sodium and vanadium, very close to those reported (Wilkinson).

Figure 3.4a, b representing the error analysis of the data used for the present analysis conform to the quality of good data sets and justify their usage for the analysis.

The results of the multipole analysis have been given in Table 3.2.

The results of multipole refinement as given in Table 3.2 show the thermal vibration parameters, the multipole parameters and the reliability indices. The large κ' parameter of the valence part of the sodium atom (one electron) denotes

Table 3.2 Multipole parameters from MEM refinement

Parameter	Na	V
Pc	9.941(364)	17.876(403)
Pv	1.132	4.573
κ'	1.691(257)	1.108(242)
κ''	1.066(187)	1.016(693)
B($\mathring{A}^2$)	8.811(713)	0.406(048)
R(%)	2.39	0.95
wR(%)	6.08	1.15
GoF	3.24	2.90

B Debye–Waller factor; Pc population coefficients in the core region

Pv population coefficients in the valence region; κ' and κ'' variable parameters

the contraction needed to restrict the lone valence electron to its nearly ideal value of one. In this work, the number of core and valence electrons has also been refined in the multipole formalism, for both sodium and vanadium, which leads to very close and reasonable values. For sodium, the core and valence electrons are refined to be 9.94 and 1.13 e$^-$ the total being 11.07 e$^-$ well within the experimental error limits. Only a large κ' parameter will restrict the large thermal vibration of sodium atom to regain this amount of electrons in the valence region. In vanadium, the value of κ' is 1.11, again showing contraction of the atom, but not as much as in sodium. The number of core and valence electrons in vanadium is found to be 17.876 and 4.573 e$^-$. There is a deficit of around 0.551 e$^-$ in vanadium, which actually shows up as a share in the bonding electron and shoots up the mid-bond density. In sodium there is no deficit in the total number of electrons leading to meagre share to the bonding and a decrease in the mid-bond density value. Figure 3.2a, b clearly show these features; clear and density-free regions in sodium and charge accumulated mid-bond regions in vanadium. Thus MEM and multipole results are complementary to each other.

The static multipole deformation maps are shown in Fig. 3.5a, b for sodium and vanadium respectively on the (100) plane. The static multipole deformation maps are shown in Fig. 3.6a, b for sodium and vanadium respectively on the ($\frac{1}{2}$00) plane.

Figure 3.5a shows positive electron contours at the centre of the (100) plane, and at the edge centres and corners, although the values of these densities are very small. Yet, right from the core, the positive densities are seen showing enhanced thermal vibration of sodium atom and the contracted spherical core in line with the κ' parameter of sodium.

The ($\frac{1}{2}$00) section (Fig. 3.6a) also shows the sodium atom with large thermal vibration of the valence region. Figure 3.5b shows the nullified core and residual valence region (in terms of thermal vibration) of vanadium atom on (100) plane. The core is not seen to be as much contracted as in the case of sodium as seen from Fig. 3.6b.

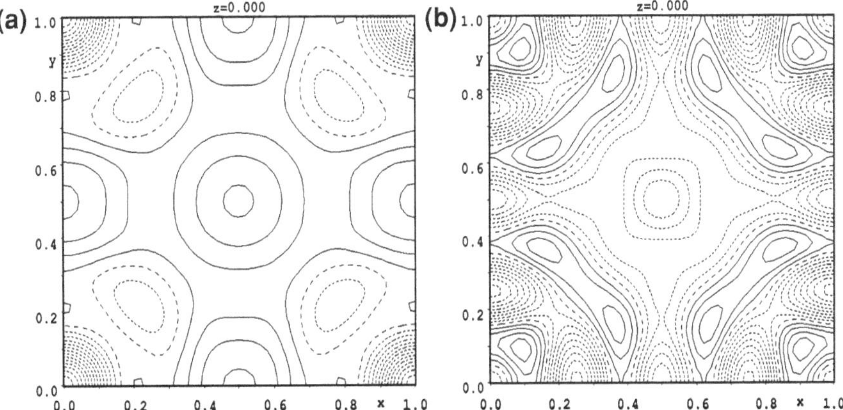

Fig. 3.5 a Static multipole deformation map of sodium on the (100) plane. *Dotted lines* indicate negative electron densities. **b** Static multipole deformation map of vanadium on the (100) plane. *Dotted lines* indicate negative electron densities

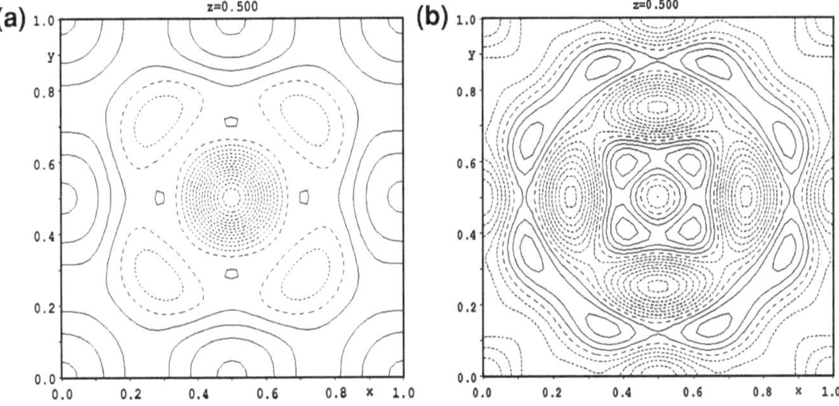

Fig. 3.6 a Static multipole deformation map of sodium on the (½00) plane. *Dotted lines* indicate negative electron densities. **b** Static multipole deformation map of vanadium on the (½00) plane. *Dotted lines* indicate negative electron densities

3.1.6 Conclusion

The nature of bonding and the charge distribution in sodium and vanadium metals has been analysed using the reported X-ray data of these metals. MEM and the multipole analysis have been used for this purpose. The bonding in these metals has been elucidated and analysed. The nature of bonding are clearly revealed by the two-dimensional MEM maps plotted on (100) and (110) planes and the mid-bond densities are clearly revealed by the one-dimensional electron density along the [100], [110] and [111] directions. The mid-bond densities in sodium and vanadium are found to be 0.014 and 0.723 e/Å^3 respectively, giving an indication of the strength of the bonds in these materials. The sodium atom is found to contract more than the vanadium atom from the multipole analysis.

3.2 Aluminium, Nickel and Copper

3.2.1 Introduction

Technological evolutions result in efforts for developing new and sophisticated materials of immense use in domestic, technical and industrial applications. Usually, the synthesis of new materials results in single-phase materials, though often not in single crystalline form. Hence, a complete analysis of the structure, local distribution of atoms and electron distribution in core, valence and bonding regions is necessary using powder diffraction methods, since most of the recent materials are initially obtained in powder form. Since one can make efforts to grow single crystals from powders, a prior analysis is required using powders to proceed for single crystal growth.

In this context, we have taken three simple metals Al, Ni and Cu and collected powder data sets to study the structure in terms of the local and average structural properties using pair distribution function (hereafter PDF), the electron density distribution between atoms using MEM and the bonding of core and valence electron distribution using multipole technique. Particularly, the PDF analysis requires data sets of very high values of Q (= 4.лsin θ/λ) which is achievable only through synchrotron studies and is not always accessible. But this work gives reasonable results, which could be obtained through single crystal work or through high Q data sets, using only powder samples with laboratory experiments. Also, a study on the electronic structure of the metals using the most versatile techniques like MEM (Collins 1982) and multipole method (Hansen and Coppens 1978) available today is carried out. If the tools available for the analysis yield highly precise information, it is appropriate to apply them to the precise available data sets, as has been done in this work, thereby testing the methodology.

In order to elucidate the distribution of valence electrons, and the contraction/expansion of atomic shells, multipole analysis of the electron densities is carried out using the software package JANA 2000 (Petříček et al. 2000). Recently, the multipole analysis of the charge densities and bonding has been widely used to study the electronic structure of materials (Coppens and Volkov 2004; Poulsen et al. 2004; Frils et al. 2004; Pillet et al. 2004; Marabello et al. 2004). In this work, the multipole model proposed by Hansen and Coppens (1978) is used for elucidating the electronic structure.

3.2.2 Summary of the Work

The average and local structures of simple metals Al, Ni and Cu have been elucidated for the first time using MEM, multipole and PDF. The bonding between constituent atoms in all the above systems is found to be well pronounced and is clearly seen from the electron density maps. The MEM maps of all the three systems show the spherical core nature of atoms. The mid-bond electron density profiles of Al, Ni and Cu reveal the metallic bonding nature. The local structure using PDF profile of Ni has been compared with that of the reported results. The R value in this work using low Q XRD data for the PDF analysis of Ni is close to the value reported using high Q synchrotron data. The cell parameters and displacement parameters were also studied and compared with the reported values.

3.2.3 Data Collection and Structural Refinement

The powder X-ray intensity data was collected at the Regional Research Laboratory (RRL), Council of Scientific and Industrial Research (CSIR), Thiruvananthapuram, India, using X-PERT PRO (Philips, Netherlands) X-ray diffractometer with a monochromatic incident beam, which offers pure Cu $K\alpha_1$ radiation. The wavelength used for X-ray intensity data collection is 0.154056 nm with 2θ range of data collection from 10 to 120°. The raw intensities were refined using the software programme JANA 2000 (Petříček et al. 2000), considering the fcc unit cell of Al, Ni and Cu with 4 atoms/unit cell in the space group Fm3m. The fitted profiles of observed and calculated relative intensities with the Bragg peaks along with the difference are given in Fig. 3.7a, b, c for Al, Ni and Cu respectively.

The results of the refinements are given in Table 3.3.

The refined structure factors are given in Table 3.4a, b, c respectively, for Al, Ni and Cu.

3.2.3.1 MEM Refinements

The refined structure factors were used for MEM analysis using the methodologies discussed in earlier works, e.g., Saravanan et al. (2003), Israel et al. (2003, 2004). In the present calculation, the unit cell was divided into 64^3 pixels and the initial electron density at each pixel was fixed uniformly as Z/a_0^3, where Z is the number of electrons in the unit cell. The electron density is evaluated by carefully selecting the Lagrange multiplier in each case such that the convergence criterion C becomes unity after performing the minimum number of iterations. The resulting parameters of MEM computations have been given in Table 3.5.

The electron density distribution of Al, Ni and Cu has been mapped using the MEM electron density values obtained through refinements. Figure 3.8a, b, c show

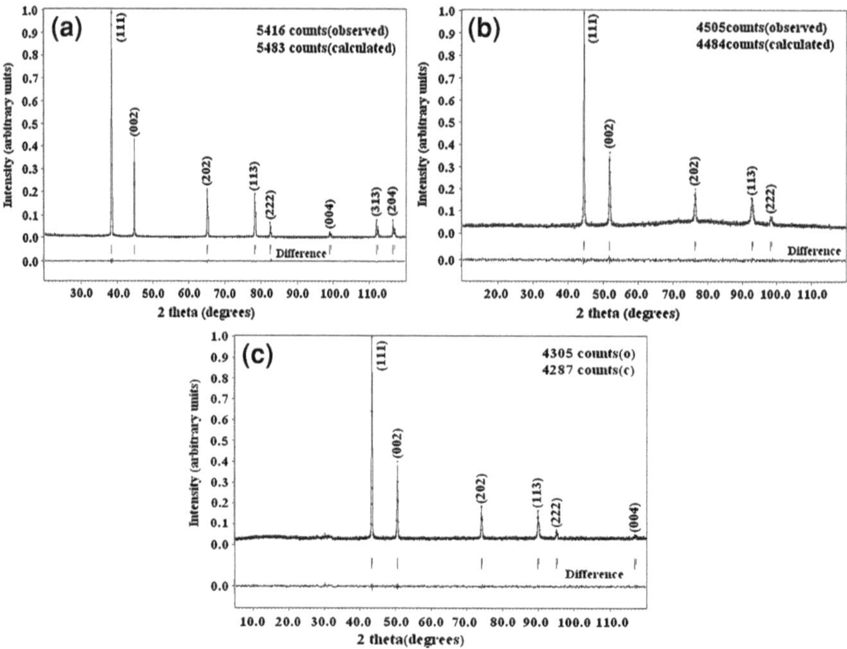

Fig. 3.7 **a** The Rietveld refined powder profile for Al along with difference between observed and calculated relative intensities. **b** The Rietveld refined powder profile for Ni along with difference between observed and calculated relative intensities. **c** The Rietveld refined powder profile for Cu along with difference between observed and calculated relative intensities

Table 3.3 The Refined structural parameters using JANA 2000

Parameters	Al	Ni	Cu
Cell parameter (Å)	4.05210(4)	3.52660(7)	3.62540(5)
Cell parameter (Å)*	4.0495	3.5238	3.6149
B (Å²)	0.983(80)	0.561(64)	0.432(53)
B (Å²)**	0.746	0.3401	0.526
$R_P\%$	9.84	6.21	6.77
$wR_P\%$	6.23	7.88	8.63

B Debye–Waller factor

the MEM electron density distribution of Al, Ni and Cu on the (100) plane of the unit cell.

Figure 3.8d, e, f show the MEM electron density distribution of Al, Ni and Cu on the (110) plane.

Figure 3.9a, b, c show the one-dimensional variation of the electron densities of Al, Ni and Cu along the three directions [100], [110] and [111] of the unit cell respectively.

The electron densities at the saddle points along these three directions have been given in Table 3.6.

Table 3.4 Observed and calculated structure factors for **a** Aluminium **b** Nickel **c** Copper

hkl	F_o	F_c	$F_o - F_c$	$\sigma(F_o)$
a Aluminium				
111	35.07	35.05	0.03	0.22
0 02	32.87	32.84	0.03	0.16
202	26.63	26.82	−0.19	0.36
113	22.87	23.18	−0.32	0.11
222	22.34	22.17	0.17	0.58
004	18.83	18.64	0.19	0.83
313	15.96	16.47	−0.51	0.22
204	16.23	15.82	0.41	0.37
b Nickel				
111	68.16	67.93	0.23	0.42
002	61.40	61.96	−0.57	0.59
202	44.38	45.76	−1.39	0.68
113	38.60	37.59	1.00	0.56
222	34.01	35.38	−1.36	1.31
c Copper				
111	78.59	78.49	0.10	0.53
002	72.71	72.65	0.06	0.76
202	55.34	55.46	−0.12	0.87
113	47.00	46.73	0.27	0.75
222	44.06	44.33	−0.27	1.67
004	35.95	36.58	−0.64	2.99

Table 3.5 Parameters of the MEM refinement

System	Prior ED $(10^{-3}$ e/Å$^3)$	Resolution (Å/Pixel)	λ	R_{MEM} (%)	wR_{MEM} (%)	Number of cycles
Al	0.7806	0.0633	0.010	1.3839	0.7897	1685
Ni	2.5538	0.0551	0.015	1.2011	1.0346	830
Cu	2.4344	0.0567	0.005	1.7610	1.2318	1072

ED Electron density, λ Lagrangean multiplier

Table 3.6 Electron density profiles along the three directions of the unit cells

Direction metals	[100]		[110]		[111]	
	Position (Å)	ED (e/Å^3)	Position (Å)	ED (e/Å^3)	Position (Å)	ED (e/Å^3)
Al	2.026	0.1467	1.433	0.2636	3.509	0.1467
Ni	1.763	0.2296	1.247	1.1010	3.054	0.2296
Cu	1.813	0.3318	1.282	0.7649	3.139	0.3317

ED Electron density

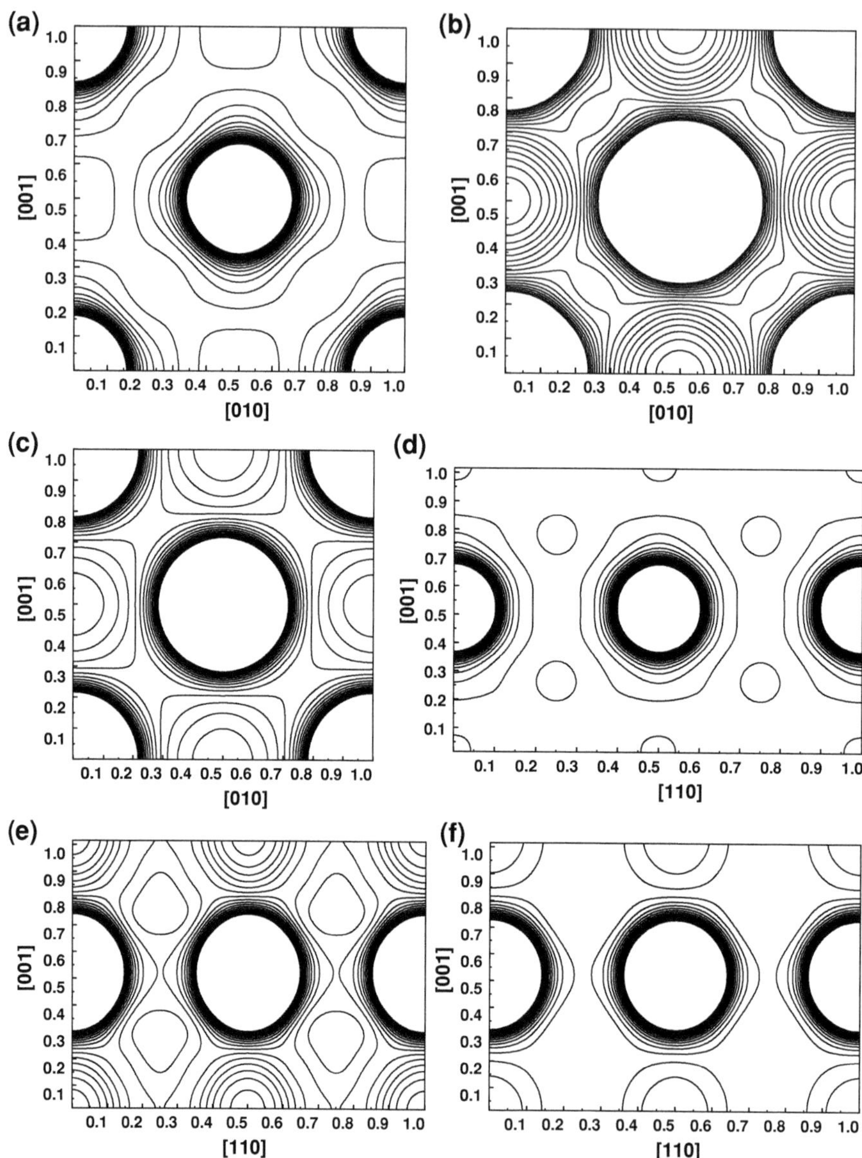

Fig. 3.8 a MEM electron density distribution of Al on the (100) plane. Contour range is 0.0–2.0 e/Å^3 and contour interval is 0.1 e/Å^3. **b** MEM electron density distribution of Ni on the (100) plane. Contour range is 0.0–2.0 e/Å^3 and contour interval is 0.1 e/Å^3. **c** MEM electron density distribution of Cu on the (100) plane. Contour range is 0.0–2.0 e/Å^3 and contour interval is 0.1 e/Å^3. **d** MEM electron density distribution of Al on the (110) plane. Contour range is 0.0 to 3.0 e/Å^3 and contour interval is 0.15 e/Å^3. **e** MEM electron density distribution of Ni on the (110) plane. Contour range is 0.0–3.0 e/Å^3 and contour interval is 0.15 e/Å^3. **f** MEM electron density distribution of Cu on the (110) plane. Contour range is 0.0–3.0 e/Å^3 and contour interval is 0.15 e/Å^3

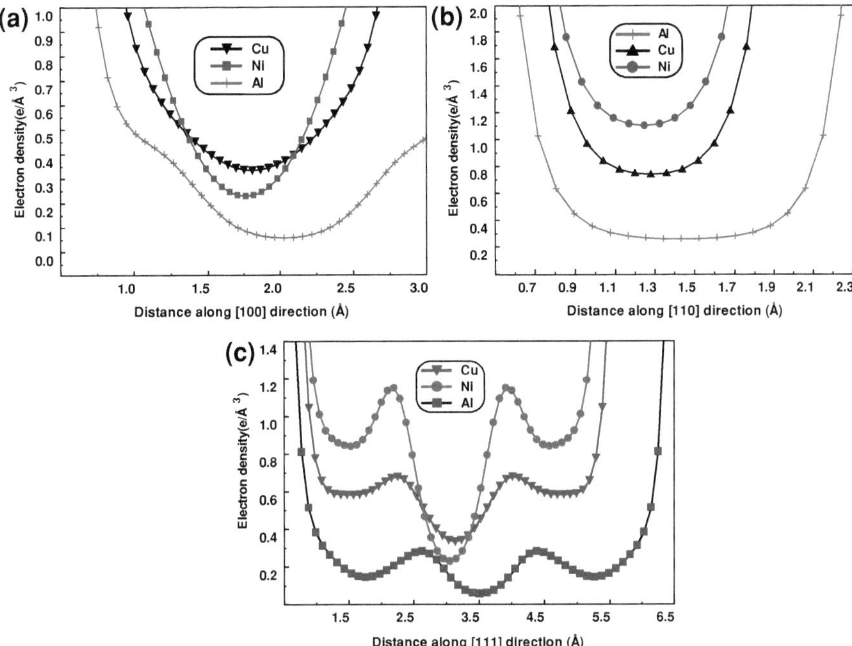

Fig. 3.9   a One-dimensional electron density profiles of Al, Ni and Cu along the [100] direction
of the unit cell. b One-dimensional electron density profiles of Al, Ni and Cu along the [110]
direction of the unit cell. c One-dimensional electron density profiles of Al, Ni and Cu along the
[111] direction of the unit cell

A detailed discussion is formulated and given in the following section based on
these two-dimensional MEM maps and one-dimensional profiles.

3.2.3.2 PDF Refinements

For materials whose structure is not reflected in the long-range order of the crystal,
an alternate structural analysis called the Pair Distribution Function (PDF)
approach is used. This method is sometimes called the real-space structure
determination method, because the PDF is modelled in real-space, rather than in
the reciprocal space. The PDF reflects the short-range ordering in a material or the
probability to find an atom at a distance r from another atom. In other words PDF
gives the bond length distribution of the material under consideration (Proffen and
Neder 1997). It well with the inter atomic distances computed from a crystallographic model, when there are no short-range deviations from the average structure (Tobi and Egami 1992; Peterson et al. 2003).

This approach has been widely used for studying the structure of glasses and
liquids (Debay and Menki 1930; Bowran and Finney 2003). More recently, it has
been applied to disordered crystalline and partially crystallised materials.

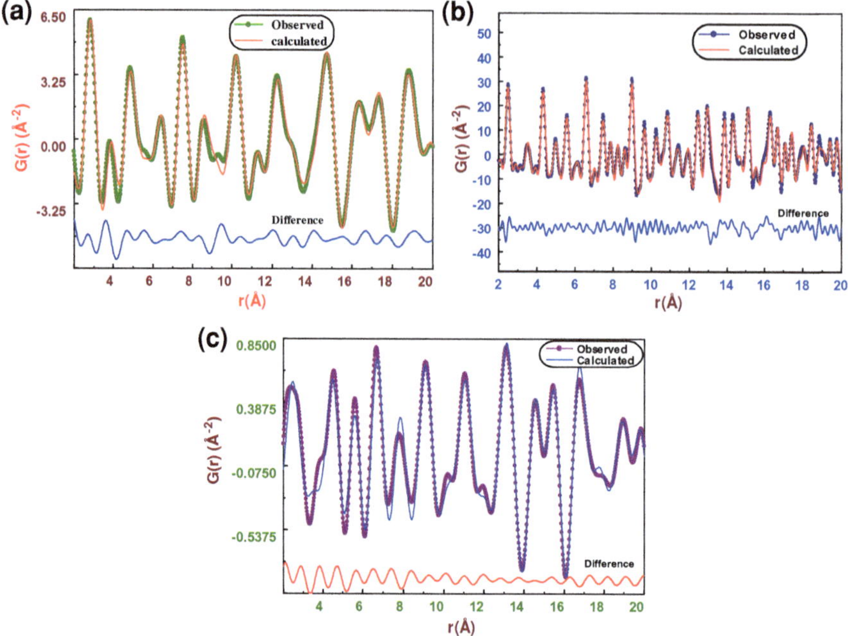

Fig. 3.10 **a** The observed and calculated PDFs of Al with difference profile. **b** The observed and calculated PDFs of Ni with difference profile. **c** The observed and calculated PDFs of Cu with difference profile

Quantitative structural information about nanometer length scales can be obtained by fitting a model directly to the PDF (Proffen and Billinge 1999) based on the equation,

$$G(r) = 4\pi r[\rho(r) - \rho_0] = \frac{2}{\pi} \int \vec{Q} \left[S\left(\vec{Q}\right) - 1 \right] \sin\left(\vec{Q}\,r\right) \mathrm{d}\vec{Q} \qquad (3.4)$$

where $G(r)$ is the atomic pair distribution function and $\rho(r)$ corresponds to the (atomic) number density at a distance r from the average atom. The atomic pair distribution function thus obtained from powder diffraction data is a valuable tool for the study of the local atomic arrangements in a material, since both Bragg and diffuse scattering information about local arrangements are preserved in PDF.

In this study, the observed PDFs of Al, Ni and Cu were obtained from the raw intensity data using a software programme PDFgetX (Jeong et al. 2001) after performing corrections such as multiple scattering correction, polarisation correction, absorption correction, normalisation correction and Compton correction. A comparison between the observed and calculated PDF has been carried out using the software package PDFFIT (Proffen and Billinge 1999). Figure 3.10a–c give the

Table 3.7 Distances from the PDF profiles

System	First nearest neighbour distance (Å)	
	Observed	Calculated[a]
Al	2.86	2.863
Ni	2.50	2.492
Cu	2.38	2.564

[a] Wyckoff 1963

fitted observed and calculated PDF profiles along with the difference between them for Al, Ni and Cu respectively, and Table 3.7 gives the bond length distribution of atoms.

3.2.3.3 Multipole Refinements

The multipole model represents an extrapolation from a finite set of experimental data. An important feature of the multipole model is the possibility to adjust the radial dependency for each atom type by including the expansion/contraction controlling κ' parameters in the refinement, and possibly also the κ'' parameters, to change the radial dependency of the valence deformation density.

In our study, we have used a modified electron density model proposed by Hansen and Coppens (1978) with the option that allows the refinement of population parameters at various orbital levels, multipoles up to seventh order and corresponding radial expansion (κ') parameters. The κ' parameters were refined in separate cycles together only with the scale factor, and convergence was achieved in an iterative process. Each refinement cycle was considered successful at the point at which the maximum shift/s.u. was less than 0.001. The parameters and results obtained from multipole refinements are given in Table 3.8.

The effect of the temperature can be distinguished from the convolution and deconvolution form of thermal contribution to the charge density as dynamic (DMD) and static multipole deformation (SMD) maps. The SMD maps offer a chance to compare the electron densities without the disturbance of the thermal vibration of atoms. Hence, the SMD maps have been computed and presented in Fig. 3.11a, b, c for Al, Ni and Cu respectively

3.2.4 Results and Discussion

The fitted profiles of Rietveld refinements (Fig. 3.7a, b, c for Al, Ni and Cu respectively) give a clear picture of the quality of the sample as well as the data. The accuracy of refinements can be judged from the difference curve between the observed and calculated intensity, which is almost a straight line. The results of the

Table 3.8 Parameters from multipole refinement

Parameters	Al	Ni	Cu
κ'	1.0069(0521)	1.1132(0136)	1.0648(3447)
B (Å^2)	0.9831(0809)	0.5604(0640)	0.4317(0533)
R (%)	0.84	2.12	0.59
wR (%)	1.2	1.61	0.36
GoF	1.16	1.16	1.09
ρ_{max} (SMD)	0.05	0.02	0.02
ρ_{min} (SMD)	−0.02	−0.08	−0.01

B Debye–Waller factor; κ' variable parameter

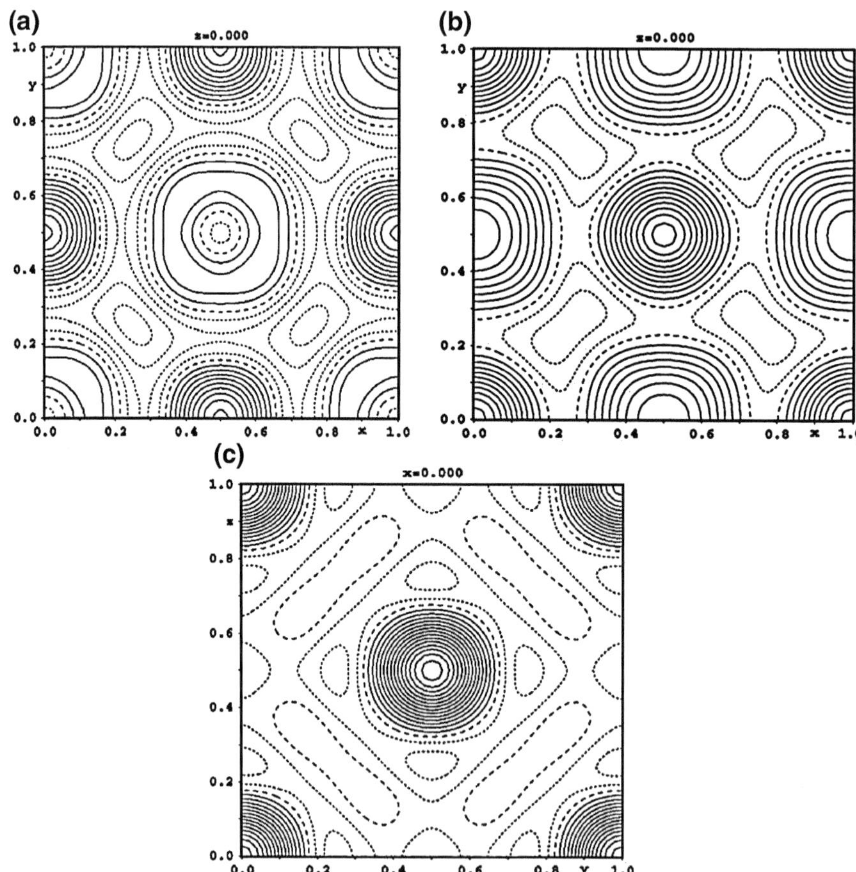

Fig. 3.11 **a** Static Multipole Deformation map of Al on (100) plane. *Dotted lines* indicate negative electron densities. **b** Static Multipole Deformation map of Ni on (100) plane. *Dotted lines* indicate negative electron densities. **c** Static Multipole Deformation map of Cu on (100) plane. *Dotted lines* indicate negative electron densities

structural refinement (Petříček et al. 2000) given in Table 3.3 show reasonable values of the Debye–Waller factor for Al, Ni and Cu obtainable from a powder data set, which is comparable to those reported in Peng et al. (1996). The experimental cell constant of Al, Ni and Cu are highly comparable to the reported values (Wyckoff 1963) and the differences are 0.0026, 0.0028 and 0.0096 Å for Al, Ni and Cu respectively, which are very low indicating the accuracy of refinements and also the precision of the observed data sets. The reliability indices for all the powder samples are very low indicating the correctness of the refinements (R_P = 9.84, 6.21 and 6.77 for Al, Ni and Cu respectively).

The MEM electron density distribution map of Al as given in Fig. 3.8a for (100) plane reveals the core of the Al atom being spherical and this sphericity persists even at slightly larger distances away from the centre. The distribution of charges all over the plane, indicates the distribution of electrons and the charges at the edge centres on (100) plane show the distribution from the perpendicular face-centred atoms, indicating the extension of the spread of the charges. The electron density map of Al on (110) plane shows (Fig. 3.8b) highly concentrated charges at the core of the atoms, and the distribution of charges on places other than the atomic positions, due to the valence, free electrons.

The core of Ni and Cu atom as seen from the electron density map on (100) plane (Fig. 3.8b, c), show the perfect spherical nature of the electronic charge clouds. The edge centres show the distribution of electron density of the atoms located at the face centres of the planes perpendicular to the paper. Since the atomic numbers of nickel and copper are higher than that of aluminium, more diffuse distribution of charges are seen in the electron density maps of Ni and Cu on the (100) plane. The electron density map of Ni and Cu on the (110) plane (Fig. 3.8e, f), show similar trends as observed in (100) plane.

The one-dimensional profiles of electron density constructed along [100], [110] and [111] directions for Al, Ni and Cu are shown in Fig. 3.9a, b and c respectively. The positions of minimum electron densities and density values have been given in Table 3.6. The mid-bond density for Al is found to be 0.263 e/Å^3 at a distance of 1.433 Å along the bonding direction [110]. The mid-bond electron density between the Al atoms along [100] direction is 0.147 e/Å^3 at a distance of 2.026 Å. The first minimum and the mid-bond electron density between the atoms along [111] direction are 0.119 e/Å^3, 0.147 e/Å^3 at a distance of 1.755 and 3.509 Å respectively.

The mid-bond density for Ni is found to be 1.101 e/Å^3 at a distance of 1.247 Å along the bonding direction [110]. Table 3.6 shows that the mid-bond density along [100] direction of Ni occurs at a distance 1.763 Å with a value 0.229 e/Å^3. The first and second minimum electron densities between the atoms along [111] directions are 0.841 and 0.229 e/Å^3 at a distance of 1.527 and 3.054 Å.

The mid-bond density for Cu is found to be 0.765 e/Å^3 at a distance of 1.282 Å along the bonding direction [110]. The mid-bond density along [100] direction occurs at a distance 1.813 Å with a value 0.332 e/Å^3. The first and second minimum electron densities between the atoms along [111] direction are 0.664 and 0.332 e/Å^3 at a distance of 1.374 and 3.139 Å respectively. The mid-bond density in Ni is the largest value (Table 3.6) among the three metal systems we have studied. But along the [100] and [111] directions (non-bonding), the element with the higher atomic number has the higher electron density. Hence obviously the bonding interaction is stronger in Ni as evidenced by the higher electron density along [110] direction.

Hence, from the qualitative and quantitative analysis of the MEM electron densities, the bonding in Al, Ni and Cu is predicted to be predominantly metallic and the interaction of charges along other nonb-onding directions for Ni and Cu seems to be heavier due to their metallic nature. (Because, if the systems are ionic one cannot expect higher mid-bond values as in the present cases and if the systems are covalent, one can expect values higher than the present values due to overlapping of charges). Also the strength of the electron density is higher in the case of Ni, which has lower number of electrons in the unit cell compared to Cu. The electron density of Al in all the three directions [100], [110] and [111] are relatively low compared to Ni and Cu, which conforms to the loosely packed structure of Al. (The interaction of atomic charges will be less and hence electron densities along bonding and directions other than bonding are expected to be minimum).

The refined pair distribution function (the Fourier transform of $S\left(\vec{Q}\right)$, the reduced structure factor) given in Fig. 3.10a, b, c for Al, Ni and Cu show nice matching of observed and calculated PDFs. The PDF refinement can be considered as equivalent to matching the observed X-ray powder data with a model with too little structure parameters. Moreover, high Q data are required for this type of analysis on short-range order and local structure of materials (Egami 1990). In spite of these factors, an attempt has been made to use the X-ray powder data for this analysis, to test how well these powder data sets can be used for purposes like PDF analysis.

The peaks in the PDF profile correspond to the nearest neighbour distances given in Table 3.7. In Al, the difference between observed and calculated first nearest neighbour distances turns out to be 0.003 Å. In Ni, the difference is 0.008 Å and for Cu it turns out to be 0.184 Å. These differences are not too high considering the fact that we have used only (i) powder data with (ii) limited Q values. There may also be local undulations in the structure of these three systems which lead to these differences. The local disorder is expected to affect the repeat distances, because of the diffuse scattering, which does not have sharp, single pointed X-ray diffraction phenomenon. Hence, it is reflected in the local structural analysis using suitable tools like PDF. From these differences (0.003 Å for Al, 0.008 Å for Ni and 0.184 Å for Cu), one can conclude that (i) the local disorder increases as the atomic number increases due to more concentration of charges and (ii) the disorder propagates at a longer distance from the origin leading to enhanced differences in neighbour distances. In the case of Cu, an additional disorder is possible due to the fact that CuKα radiation has been used for the powder data collection which gives rise to resonance effects and possible fluorescence.

The PDF profile of Ni has been compared with that obtained with reported data (Proffen and Billinge 1999). Very close resemblance is seen between these two refinement results. The R values are also close to each other [18% in this work compared to 14% in the work by Proffen and Billinge (1999)].

The multipole refinement shows no expansion/contraction for Al as seen from the κ' value given in Table 3.8 [$\kappa' = 1.007(0.052)$]. In the case of Cu, there is a slight contraction of the atom, but in Ni there is appreciable contraction compared to the other two systems. There is a possibility that the contraction is effected to obey Pauli's principle and to make the system with the strongest mid-bond interaction stable (as evidenced by the mid-bond electron density of Ni).

The results of SMD calculations given in Table 3.8 show the positive and negative (maximum and minimum) electron densities for all the three systems. Since these calculations are based on difference densities, the lowest electron densities are seen for all the systems, i.e., the maximum positive difference is 0.05 e/Å^3 for Al and the maximum negative density is -0.08 e/Å^3 for Ni. These values indicate the accuracy with which the multipole analysis has been carried out. Moreover, the positive difference densities are the same for both Ni and Cu (0.02 e/Å^3), thereby justifying the fact that no biassed results are obtained as far as the comparison of electron densities of Ni and Cu is concerned.

3.2.5 Conclusion

A clear understanding of the structure of the metals A1, Ni and Cu has been gained from this work, in terms of cell parameters, thermal vibrations and electron density distribution in the bonding region using MEM and local structural analysis using PDF. Multipole model analysis reveals the clear qualitative and quantitative picture of the charge densities both in core and valance regions with the expansion and contraction of the valence region.

3.3 Magnesium, Titanium, Iron, Zinc, Tin and Tellurium

3.3.1 Introduction

The applications of metallic materials are spread into various areas of science and technology including electronics, aeronautics and space travel, to name a few. Hence a study on some technologically important metals is essential in terms of the local and average structures which are completely different. The usual methods of analysis using structural refinement of X-ray or neutron data will give only the average structure of the materials under investigation. There is only limited information available about the investigations of materials in terms of the local structure. In the present investigation, powder data sets of six elemental metals magnesium, titanium, iron, zinc, tin and tellurium were collected to study the structure in terms of the electron density distribution between atoms using MEM and the local and average structural properties using PDF. The PDF analysis requires data sets of very high values of Q (= $4\pi \sin \theta/\lambda$) which is achievable only through synchrotron studies, which is not always accessible. But this work gives reasonable results, which could be obtained through high Q data sets.

3.3.2 Summary of the Work

The average and local structures of the metals magnesium, titanium, iron, zinc, tin and tellurium have been analysed using MEM, and PDF. The bonding between the constituent atoms in all these systems is found to be well pronounced and is clearly seen from the electron density maps. The mid-bond electron density profiles reveal the metallic nature of the bonding. The local structure of Mg, Ti, Fe, Zn, Sn and Te using the low Q XRD have given results comparable to high Q synchrotron data. The cell and displacement parameters are also studied.

3.3.3 Maximum Entropy Method

For the accurate charge density study, the MEM (Sakata and Sato 1990; Collins 1982) has been used. It is a non-linear calculation that has its roots in information and probability theory, and was originally designed to reconstruct the most probable and least biassed probability distribution in an underdetermined situation. The MEM can yield a high-resolution density distribution from a limited number of diffraction data without using a structural model. The obtained density distribution gives detailed structural information in the inter atomic region, e.g. bonding electron densities. The ability of MEM in terms of a model-free reconstruction of the charge densities from measured X-ray diffraction data can be interpreted as 'imaging of diffraction data'. Combining MEM with the Rietveld method (Rietveld 1969) has been successful in developing a sophisticated method of structure refinement in charge density level, that is, the MEM/Rietveld (Takata et al. 1995). The MEM/Rietveld analysis is an iterative method combining the MEM and Rietveld analyses. The MEM charge density at certain iteration step normally succeeds in providing a better structural model for the Rietveld of the next iteration. In this method, the final MEM electron density distribution derived is compatible with the structural model used in Rietveld refinement. The method has been successfully applied to the structural studies of metals and alloys (Robert et al. 2009; Saravanan and Prema Rani 2007), semiconductors (Saravanan et al. 2008), magnetic materials (Syed Ali et al. 2009]) In this study this method has been used to elucidate the electron density distribution of elemental metals.

3.3.4 Pair Distribution Function

Conventional structure determination is based on the analysis of the intensities and positions of Bragg reflections which only allow the determination of the long-range average structure of the crystal. The key to a deeper understanding of the properties of the materials is the study of deviations from the average structure or the study of the local atomic arrangements. Deviations from the average structure result in the occurrence of diffuse scattering which contain information about two-body interactions. One method to reveal the local structure of crystals is the analysis of PDF. PDF can be understood as a bond length distribution between all pairs of atoms within the crystal (up to a maximum distance). PDF is the instantaneous atomic number density–density correlation function which describes the atomic arrangement in materials. It is the sine Fourier transform of the experimentally observable total structure factor obtained from a powder diffraction experiment. Since the total structure factor includes both the Bragg scattered intensities and the diffuse scattering part of the diffraction spectrum its Fourier associate, the PDF, yields both the local and average atomic structures of materials. Determining the PDF has been the approach of choice for characterising glasses, liquids and amorphous materials for a long time. However, its widespread application to crystalline materials, where some deviation from the average structure is expected to take place, has been relatively recent. The PDF refinement can be considered as equivalent to matching the observed X-ray powder data with a model with too little structure parameters.

Quantitative structural information about nanometer length scales can be obtained by fitting a model directly to the PDF (Proffen and Billinge 1999) based on the equation

$$G(r) = 4\pi r[\rho(r) - \rho_0]$$
$$= \frac{2}{\pi} \int \vec{Q}\left[s(\vec{Q}) - 1\right] \sin\left(\vec{Q}r\right) d\vec{Q},$$

where $G(r)$ is the atomic pair distribution function and $\rho(r)$ corresponds to the (atomic) number density at a distance r from the average atom. The atomic pair distribution function, obtained from powder diffraction data is thus a valuable tool for the study of the local atomic arrangements in a material, since both Bragg and diffuse scattering information about local arrangements are preserved in PDF.

3.3.5 Data Collection and Structural Refinement

High quality powder samples of Mg, Ti, Fe, Zn, Sn and Te were purchased and used for this work. The general properties of these metals are given in Table 3.9.

Table 3.9 General properties of the metals

Properties[a]	Mg	Ti	Fe	Zn	Sn	Te
Atomic number	12	22	26	30	50	52
Cell constant($\mathring{A}$)	a = 3.2095	a = 2.9512	a = b = c = 2.8665	a = 2.6650	a = 5.8317	a = 4.4568
	b = 3.2095	b = 2.9512		b = 2.6650	b = 5.8317	b = 4.4568
	c = 5.2107	c = 4.6845		c = 4.9470	c = 3.1324	c = 5.9270
Structure	Hexagonal	Hexagonal	Cubic	Hexagonal	Tetragonal	Hexagonal
Space group	P63/mmc	P63/mmc	Im-3 m	P63/mmc	I41/amd	P3121

[a] CRC handbook (Ram 1985)

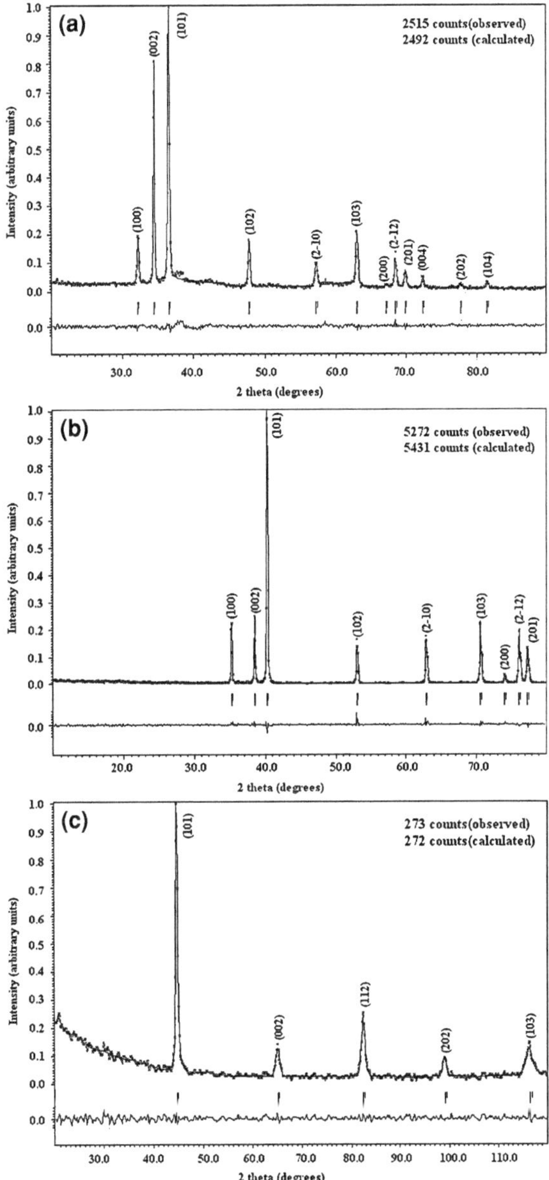

Fig. 3.12 **a** The Rietveld refined powder profile for Mg along with difference between observed and calculated relative intensities. **b** The Rietveld refined powder profile for Ti along with difference between observed and calculated relative intensities. **c** The Rietveld refined powder profile for Fe along with difference between observed and calculated relative intensities. **d** The Rietveld refined powder profile for Zn along with difference between observed and calculated relative intensities. **e** The Rietveld refined powder profile for Sn along with difference between observed and calculated relative intensities. **f** The Rietveld refined powder profile for Te along with difference between observed and calculated relative intensities

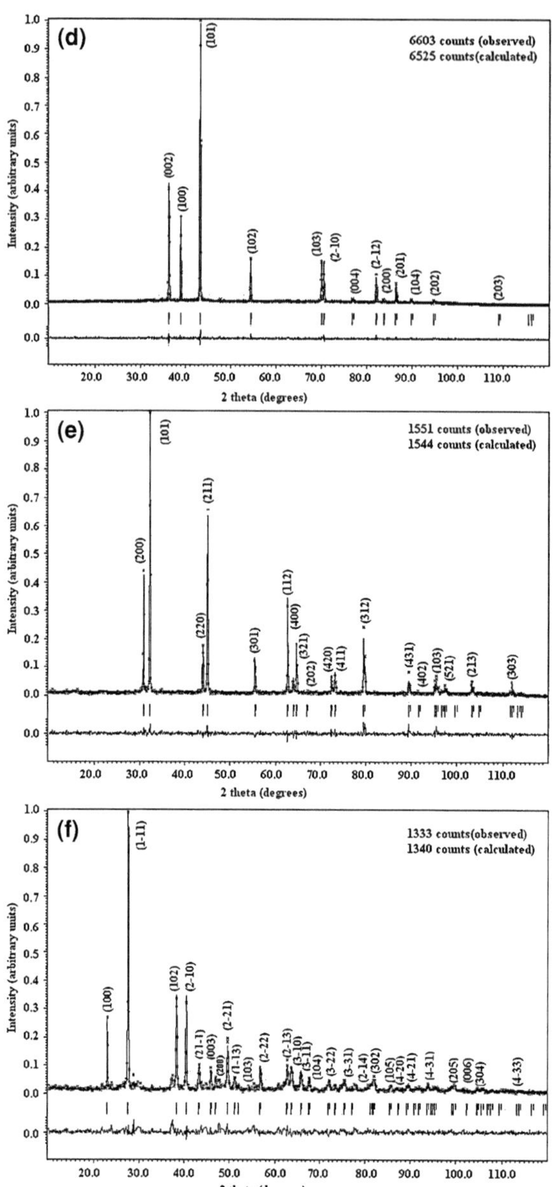

Fig. 3.12 (continued)

Table 3.10 Observed and calculated structure factors for **a** Magnesium **b** Titanium **c** Iron **d** Zinc **e** Tin **f** Tellurium

hkl	F_o	F_c	$F_o - F_c$	$\sigma(F_o)$
a Magnesium				
220	130.19	130.30	−0.11	3.41
101	109.31	109.82	−0.51	0.95
200	158.93	158.44	0.49	0.73
220	130.19	130.30	−0.11	3.41
211	93.60	93.52	0.08	1.34
301	78.01	78.25	−0.24	2.57
112	103.02	103.72	−0.70	2.13
400	95.89	101.90	−6.01	3.17
321	71.97	72.39	−0.42	2.05
420	96.46	93.20	3.26	3.03
411	64.56	62.40	2.17	2.03
312	89.12	85.19	3.93	2.15
501	65.04	54.87	10.17	3.40
431	57.19	48.25	8.94	2.99
103	51.03	47.87	3.16	2.16
332	70.87	66.37	4.50	2.76
440	69.55	64.68	4.87	2.57
521	53.62	47.79	5.83	2.20
213	40.20	43.10	−2.90	2.14
600	56.13	60.17	−4.03	2.99
303	34.93	40.05	−5.12	1.22
512	48.75	55.56	−6.81	1.64
620	47.57	54.20	−6.63	1.60
611	33.15	37.77	4.63	1.12
b Titanium				
100	8.85	8.93	−0.09	0.20
002	17.47	17.46	0.01	0.07
101	14.83	14.79	0.04	0.04
102	7.64	7.63	0.01	0.16
2−10	13.60	13.76	−0.16	0.34
103	11.00	11.17	−0.17	0.21
200	5.97	6.14	−0.17	0.61
2−12	12.51	12.10	0.41	0.30
201	10.17	10.31	0.14	0.34
004	11.44	11.56	−0.12	0.55
202	5.53	5.43	0.10	0.10
104	5.28	5.20	0.09	0.24
203	7.76	8.13	−0.36	0.41
3−10	4.52	4.49	0.02	0.85
3−11	7.66	7.57	0.09	0.35
2−14	8.08	8.52	−0.44	0.38
105	6.85	6.99	−0.15	0.28
3−12	3.97	4.03	−0.06	0.16

(continued)

Table 3.10 (continued)

hkl	F_o	F_c	F_o-F_c	$\sigma(F_o)$
204	3.29	3.88	−0.59	0.74
300	8.00	7.45	0.55	1.25
3–13	5.99	6.14	−0.15	0.34
c Iron				
101	34.32	34.31	0.01	0.82
002	25.73	26.36	−0.63	3.28
112	21.46	22.31	−0.85	2.62
202	18.57	19.02	−0.44	2.62
103	17.48	16.64	0.84	2.60
d Zinc				
002	40.22	40.08	0.14	0.31
100	20.71	20.57	0.14	0.19
101	29.06	29.06	0.00	0.13
102	13.67	13.98	−0.31	0.16
103	16.33	16.52	−0.19	0.16
2–10	20.75	21.14	−0.40	0.19
004	13.09	14.30	−1.21	0.80
2–12	14.89	14.63	0.26	0.25
200	6.34	6.08	0.26	0.59
201	12.98	13.09	−0.11	0.26
104	4.38	1.99	2.39	1.40
202	5.95	6.34	−0.40	0.36
203	2.02	2.40	−0.38	0.58
105	1.44	1.74	−0.30	0.56
2–14	1.35	1.62	−0.28	0.52
e Tin				
101	109.31	109.82	−0.51	0.95
200	158.93	158.44	0.49	0.73
220	130.19	130.30	−0.11	3.41
211	93.60	93.52	0.08	1.34
301	78.01	78.25	−0.25	2.57
112	103.02	103.72	−0.70	2.13
400	95.90	101.90	−6.01	3.17
321	71.97	72.39	−0.42	2.05
420	96.46	93.20	3.26	3.03
411	64.56	62.40	2.17	2.03
312	89.12	85.19	3.93	2.15
501	65.04	54.87	10.17	3.40
431	57.19	48.25	8.94	2.99
103	51.03	47.87	3.16	2.16
332	70.87	66.37	4.50	2.76
440	69.55	64.68	4.87	2.57
521	53.62	47.79	5.83	2.20
213	40.20	43.10	−2.90	2.14
600	56.13	60.17	−4.03	2.99

(continued)

Table 3.10 (continued)

hkl	F_o	F_c	F_o-F_c	$\sigma(F_o)$
303	34.93	40.05	−5.12	1.22
512	48.75	55.56	−6.81	1.64
620	47.57	54.20	−6.63	1.60
611	33.15	37.77	4.63	1.12
f Tellurium				
100	36.84	37.00	−0.16	0.82
1–11	27.50	27.52	−0.02	0.34
101	21.66	121.77	−0.11	1.52
102	24.56	24.73	−0.17	0.49
1–12	107.46	108.20	−0.74	2.14
2–10	92.11	92.39	−0.29	1.73
2–11	44.26	44.56	−0.30	1.68
003	108.01	104.82	3.19	5.31
200	35.72	35.56	0.16	2.55
201	56.89	55.73	1.16	1.50
2–21	79.26	77.64	1.61	2.09
2–12	37.87	38.72	−0.85	2.32
1–13	25.72	25.36	0.35	2.84
103	24.46	24.13	0.34	2.70
202	67.88	72.43	−4.55	2.60
2–22	47.88	51.09	−3.21	1.83
2–13	75.057	72.98	2.08	2.58
3–10	54.41	57.02	−2.61	1.80
3–21	37.44	38.82	−1.38	1.34
3–11	49.38	51.20	−1.82	1.77
1–14	15.67	15.20	0.46	0.73
104	79.78	77.42	2.36	3.70
2–23	26.25	25.95	0.30	1.25
203	28.45	28.45	0.33	1.37
3–12	35.00	37.31	−2.31	2.00
3–22	45.53	48.54	−3.01	2.61
300	39.75	38.33	1.43	4.28
301	30.89	27.13	3.78	1.84
3–31	69.52	61.05	8.47	4.15
2–14	25.24	25.24	−2.59	8.76
3 32	23.78	23.64	0.14	1.13
302	55.26	54.92	0.33	2.63
2–24	56.21	56.14	56.14	2.13
204	32.87	32.83	0.04	1.24
3–13	42.36	42.31	0.05	1.60
3–23	44.61	44.55	0.06	1.69
1–15	68.28	63.03	5.25	7.01
105	13.90	12.83	1.07	1.43
4–20	27.79	25.09	2.70	7.39
4–21	47.27	43.46	3.81	3.65

(continued)

Table 3.10 (continued)

hkl	F_o	F_c	$F_o - F_c$	$\sigma(F_o)$
303	34.53	35.34	−0.81	3.82
3–33	33.62	34.41	−0.79	3.72
4–10	27.16	26.50	0.66	2.36
4–11	50.46	52.94	−2.48	2.48
4–31	19.28	20.23	−0.95	0.95
2–15	17.99	19.28	−1.29	0.76
4–22	36.61	39.04	−2.44	1.62
3–14	38.03	41.12	−3.09	1.95
3–24	23.21	25.10	−1.89	1.19
2–25	27.40	27.03	0.37	1.63
205	49.03	48.37	0.66	2.92
4–12	19.90	20.61	−0.71	1.20
4–32	48.21	49.94	−1.73	2.90
006	52.05	54.31	−2.25	16.61
3-34	40.97	40.88	0.10	3.07
304	22.50	22.45	0.05	1.68
4–23	24.25	23.86	0.38	1.72
400	55.51	53.59	1.92	3.20
106	9.49	9.42	0.07	0.66
1–16	10.89	10.81	0.09	0.76
401	10.10	10.05	0.95	1.35
4–41	12.94	11.82	1.12	1.60
4–13	21.82	24.54	−2.72	4.64
4–33	18.71	21.04	−2.33	3.98
3–15	20.91	23.51	−2.60	2.18
3–25	32.03	36.01	−3.98	3.34
402	12.93	14.41	−1.48	1.32
4–42	9.29	10.35	−1.06	0.95
2–16	41.09	43.38	−2.29	3.13
4–24	29.54	31.93	−2.39	2.00
4–10	27.16	26.50	0.66	2.36
4–11	50.49	52.94	−2.48	2.48
4–31	19.28	20.23	−0.95	0.95
2–15	17.99	19.28	−1.293	0.76
4–22	36.61	39.04	−2.44	1.61
3–14	38.03	41.12	−3.09	1.95
3–24	23.21	25.10	−1.89	1.19
2–25	27.40	27.03	0.37	1.63
205	49.03	48.37	0.66	2.92
4–12	19.90	20.61	−0.71	1.20
4–32	48.21	49.94	−1.73	2.90
006	52.05	54.31	−2.25	16.61
3–34	40.97	40.88	0.10	3.07
304	22.50	22.45	0.05	1.68

Table 3.11 Refined structural parameters from JANA 2006

Parameters	Mg	Ti	Fe	Zn	Sn	Te
B (Å^2)	0.87(00)	0.55 (09)	0.55(00)	1.46(00)	2.08(00)	1.53(41)
B (Å^2)	1.46	–	0.35	1.26	–	0.74
R_{obs}(%)	1.91	1.65	2.36	1.76	4.96	3.81
wR_{obs}(%)	0.82	1.71	1.18	1.16	3.73	3.33
Rp(%)	10.03	8.92	20.89	9.08	15.84	17.11
wRp(%)	14.23	13.08	28.94	12.72	23.08	23.84
Cell parameters (Å)	a = 3.2138 (0.0011)	a = 2.954 (0.0000)	a = b = c = 2.8692 (0.00)	a = 2.6733 (0.0000)	a = 5.8022 (0.0000)	a = 4.4634 (0.0044)
	b = 3.2137 (0.0000)	b = 2.954 (0.0000)		b = 2.6733 (0.0000)	b = 5.8022 (0.0000)	b = 4.4634 (0.0000)
	c = 5.2150 (0.0019)	c = 4.6916 (0.0001)		c = 4.9652 (0.0000)	c = 3.1654 (0.0000)	c = 5.9330 (0.0058)

B reported Debye–Waller factor

Table 3.12 Refined parameters from MEM technique

Metal	Parameters				
	Number of cycles	Prior density $\tau(r_i)$ (e/Å^3)	Lagrange parameter (λ)	R_{MEM}(%)	wR_{MEM} (%)
Mg	159	0.5150	0.0026	2.5822	1.4972
Ti	45	1.2394	0.0052	1.2171	1.1493
Fe	60	2.2034	0.0327	8.5663	4.6957
Zn	205	1.9544	0.0308	2.7252	1.0814
Sn	92	1.8762	0.0423	9.7379	2.1222
Te	222	1.5220	0.0813	3.7133	3.8628

The powder X-ray intensity data were collected at the National Institute of Interdisciplinary Science and Technology (NIIST), Trivandrum, India, using X-PERT PRO (Philips, Netherlands) X-ray diffractometer with a monochromatic incident beam, which offers pure Cu Kα1 radiation. The wavelength used for X-ray intensity data collection is 0.154056 nm with 2θ range of data collection from 10 to 120° and 0.05° step size.

The structural parameters of the metals were refined with the well-known (Rietveld 1969) powder profile fitting methodology using the software package JANA 2006 (Petříček et al. 2006). The initial cell parameters and space group were fixed as given in Table 3.9. The fitted profiles and the positions of Bragg peaks for the six metals have been shown in Fig. 3.12a–f. In these figures, the dots represent the observed powder patterns and the continuous lines represent the calculated powder profiles of the respective samples. The small vertical lines indicate Bragg positions. The difference between the observed profile and fitted calculated profile is shown at the bottom of each figure. The refined profiles show good matching between the observed and calculated profiles for all the three systems.

The refined structure factors of the samples given in Tables 3.10a–f show very good agreement of the observed and calculated structure factors with very low statistical errors.

The refined structural parameters from the software JANA 2006 are given in Table 3.11. The refinement indices are low which show the accuracy of the refinement (Rp = 10.03, 8.92, 20.89, 9.08, 15.84, 17.11% respectively, for samples Mg, Ti, Fe, Zn, Sn and Te).

3.3.6 MEM Refinements

The refined structure factors were used for MEM analysis. In the present calculation, the unit cell was divided into 64^3 pixels and the initial electron density at each pixel was fixed uniformly as Z/a$_0^3$, where Z is the number of electrons in the unit cell. The electron density is evaluated by carefully selecting the Lagrange

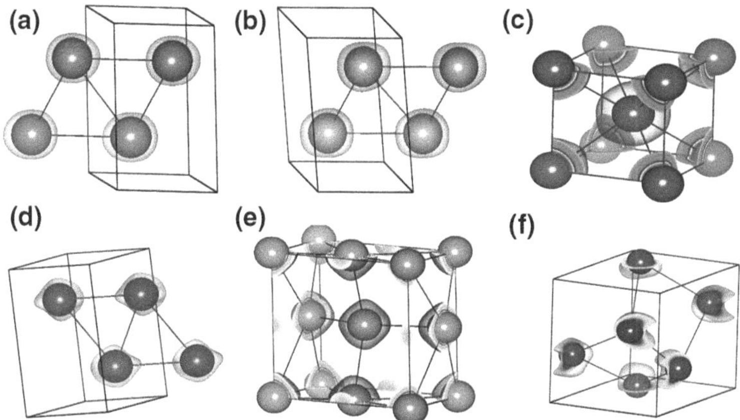

Fig. 3.13 **a** Three-dimensional MEM electron density isosurface of Mg. **b** Three-dimensional MEM electron density isosurface of Ti. **c** Three-dimensional MEM electron density isosurface of Fe. **d** Three-dimensional MEM electron density isosurface of Zn. **e** Three-dimensional MEM electron density isosurface of Sn. **f** Three-dimensional MEM electron density isosurface of Te

multiplier in each case such that the convergence criterion C becomes unity after performing minimum number of iterations. The resulting parameters of MEM computations are given in Table 3.12.

The electron density distributions of Mg, Ti, Fe, Zn, Sn and Te have been mapped using the MEM electron density values obtained through refinements. Three-dimensional MEM electron density isosurface of Mg, Ti, Fe, Zn, Sn and Te are shown in Fig. 3.13a–f respectively.

Figure 3.14a–f show the MEM electron density distributions of Mg, Ti, Fe, Zn, Sn and Te on the (100) plane of the unit cell.

Figure 3.15a–f show the MEM electron density distributions of Mg, Ti, Fe, Zn, Sn and Te on the (110) plane.

Figure 3.16a, b, c show the one-dimensional variation of the electron densities of Mg, Ti, Fe, Zn, Sn along the three directions [100], [110] and [111] of the unit cell, respectively.

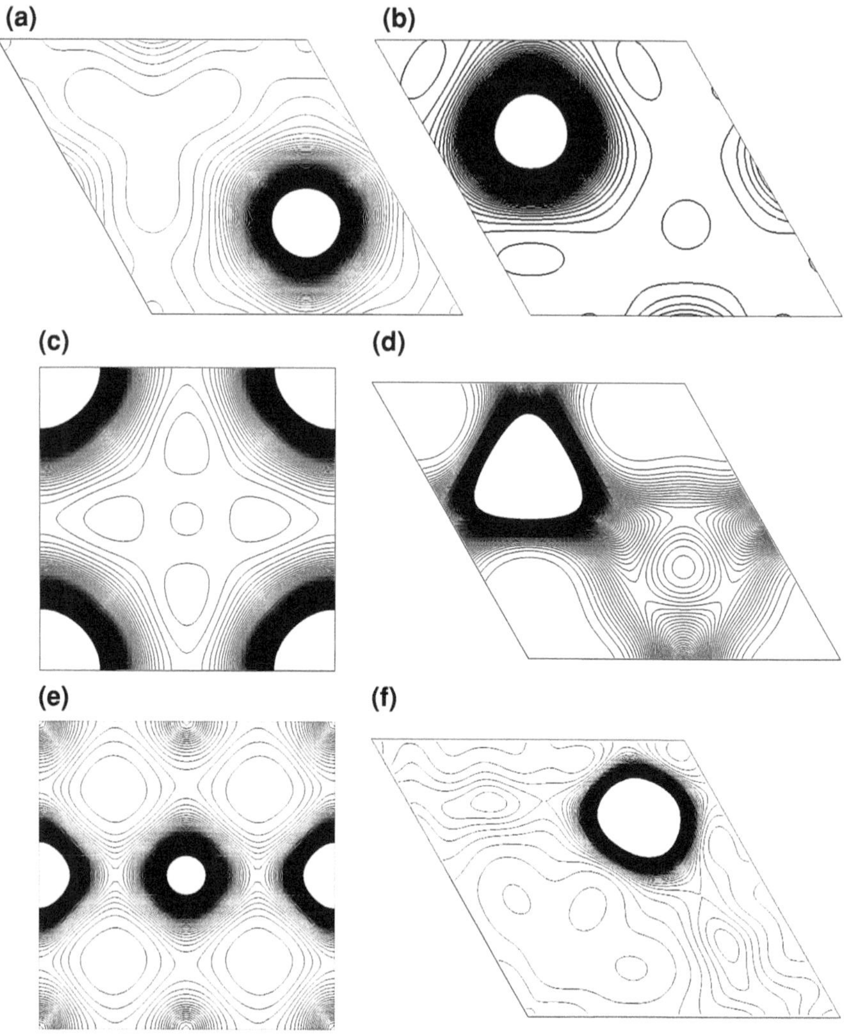

Fig. 3.14 a MEM electron density of magnesium on (100) plane. *Contour lines* are between 0.02 and 2.0 e/Å^3. Contour interval is 0.33 e/Å^3. **b** MEM electron density of titanium on the (100) plane. *Contour lines* are between 0.05 and 5.0 e/Å^3. Contour interval is 0.33 e/Å^3. **c** MEM electron density of iron on (100) plane. *Contour lines* are between 0.03 and 5.0 e/Å^3. Contour interval is 0.33 e/Å^3. **d** MEM electron density of zinc on the (100) plane. *Contour lines* are between 0.05 and 5.0 e/Å^3. Contour interval is 0.33 e/Å^3. **e** MEM electron density of tin on (100) plane. *Contour lines* are between 0.05 and 5.0 e/Å^3. Contour interval is 0.33 e/Å^3. **f** MEM electron density of tellurium on the (100) plane. *Contour lines* are between 0.05 and 5.0 e/Å^3. Contour interval is 0.001 e/Å^3

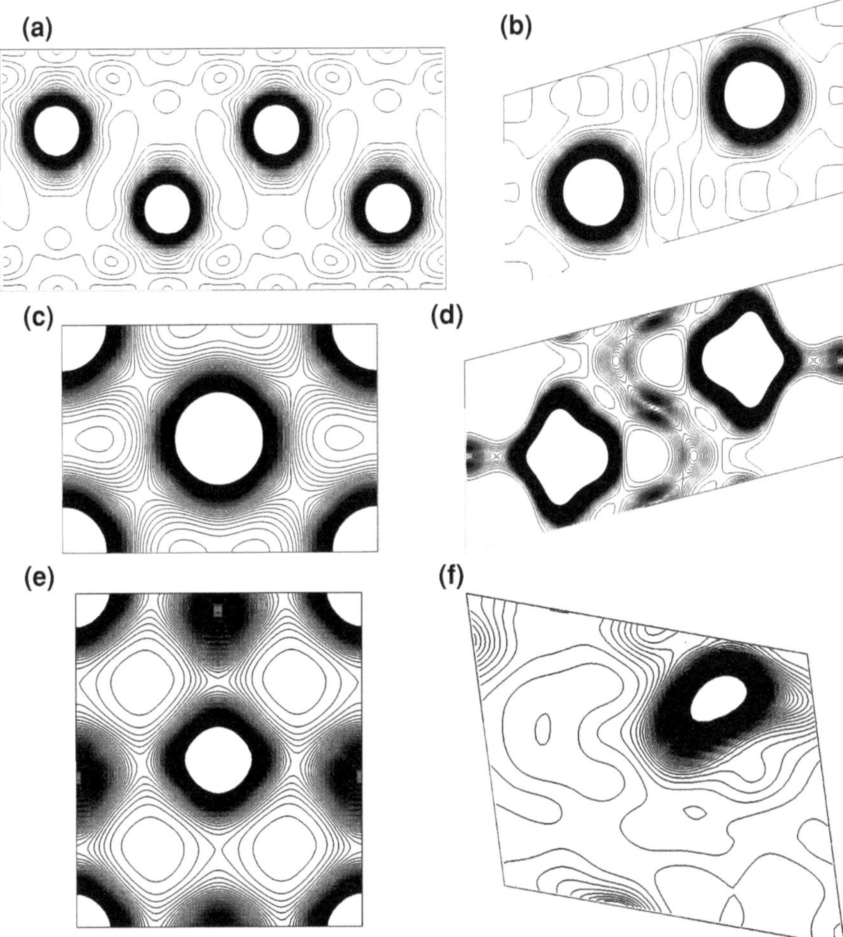

Fig. 3.15   a MEM electron density of magnesium on (110) plane. *Contour lines* are between 0.0 and 2.0 e/Å^3. Contour interval is 0.02 e/Å^3. **b** MEM electron density of titanium on the (110) plane. *Contour lines* are between 0.05 and 5.0 e/Å^3. Contour interval is 0.001 e/Å^3. **c** MEM electron density of iron on (110) plane. *Contour lines* are 0.0 and 5.0 e/Å^3. Contour interval is 0.03 e/Å^3. **d** MEM electron density of zinc on (110) plane. *Contour lines* are between 0.00 and 10.0 e/Å^3. Contour interval is 0.05 e/Å^3. **e** MEM electron density of tin on (110) plane. *Contour lines* are between 0.00 and 5.0 e/Å^3. Contour interval is 0.05 e/Å^3. **f** MEM electron density of tellurium on the (110) plane. *Contour lines* are between 0.00 and 5.0 e/Å^3. Contour interval is 0.05 e/Å^3

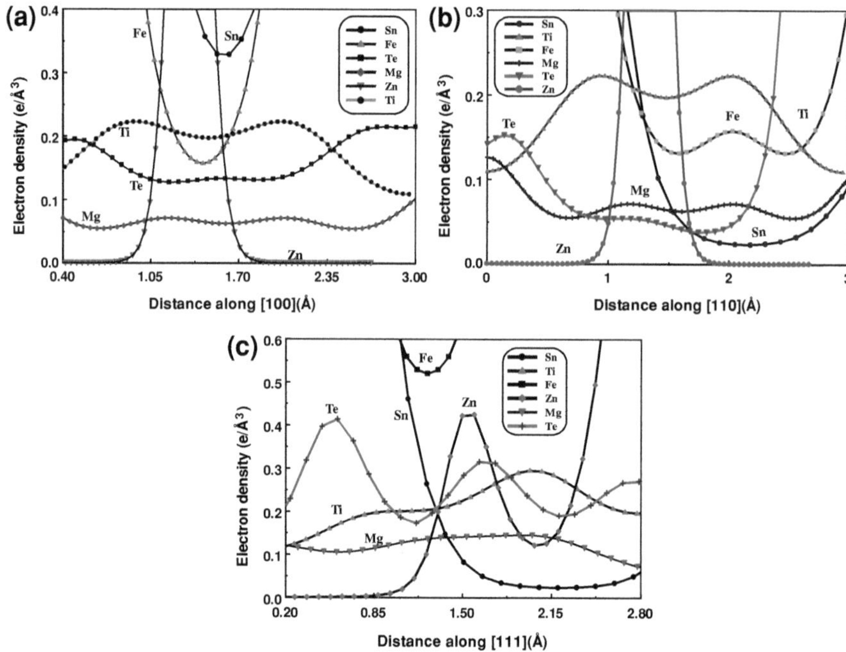

Fig. 3.16 **a** One-dimensional electron density profiles of Mg, Ti, Fe, Zn, Sn, Te along the [100] direction of the unit cell. **b** One-dimensional electron density profiles of Mg, Ti, Fe, Zn, Sn, Te along the [110] direction of the unit cell. **c** One-dimensional electron density profiles of Mg, Ti, Fe, Zn, Sn, Te along the [111] direction of the unit cell

3.3.7 Results and Discussion

3.3.7.1 MEM

The fitted profiles of Rietveld refinements (Fig. 3.12a–e for Mg, Ti, Fe, Zn, Sn and Te respectively) give a clear picture of the quality of the sample as well as the data. The accuracy of refinements can be judged from the difference curve between the observed and calculated intensity, which is almost a straight line. The results of the structural refinement (Petřiček et al. 2000) given in Table 3.10 shows reasonable values of the Debye–Waller factor for Mg (1.46 $\mathring{A}^2$) (Peng et al. 1996), Fe (0.350 $\mathring{A}^2$) (Mohanlal 1979), Zn (1.26 $\mathring{A}^2$) (Bashir et al. 1988) and Te (0.74 $\mathring{A}^2$) (Bashir et al. 1988]) obtainable from a powder data set, which is comparable to those reported. The reliability indices for all the powder samples are very low indicating the accuracy of the refinements (R_P = 10.03, 8.92, 20.89, 9.08, 15.84, 17.11% for Mg, Ti, Fe, Zn, Sn and Te, respectively).

The MEM electron density distribution map of Mg as given in Fig. 3.14a for (100) plane reveals the core of the Mg atom to be spherical and shows that the

Table 3.13 Electron density profiles along the three directions of the unit cells

Direction metal	[100]		[110]		[111]	
	Position (Å)	ED (e/Å^3)	Position (Å)	ED (e/Å^3)	Position (Å)	ED (e/Å^3)
Mg	0.6527	0.06	0.6751	0.06	0.5741	0.10
Ti	1.4772	0.19	1.4772	0.19	2.7721	0.19
Fe	1.4346	0.16	1.5850	0.13	1.2424	0.52
Zn	0.4593	0.00	0.5011	0.00	3.6121	0.12
Sn	1.6319	0.33	2.1796	0.02	2.1987	0.02
Te	1.1857	0.13	0.9764	0.05	1.1601	0.17

ED Electron density

valence region of Mg is not spherical but smearing of charges is seen. The MEM electron density distribution map of Ti as given in Fig. 3.14b for (100) plane reveals the core of the Ti atom to be spherical but the valence region of Ti shows a concave nature. The MEM electron density distribution map of Fe as given in Fig. 3.14c for (100) plane reveals the core of the Fe atom being spherical and this sphericity persists even at slightly larger distances away from the centre. The distribution of charges all over the plane indicates the distribution of electrons and the charges at the edge centres on (100) plane showing the distribution from the perpendicular face-centred atoms and indicating the extension of the spread of the charges. The MEM electron density distribution map of Zn as given in Fig. 3.14d for (100) planes reveals the distortion of the core atom which may be due to the high Debye–Waller factor as seen in Table 3.11. The distortion extends to the valence region which is also concave. The MEM electron density distribution map of Sn as given in Fig. 3.14e for (100) plane reveals the distortion in the core region which is again substantiated by the high Debye–Waller factor as seen in Table 3.11. The atomic number of Sn is higher and hence, more diffuse distribution of charges is seen in the electron density map. Due to high-density of charges, there is smearing of charges in the valence region. The MEM electron density distribution map of Te as given in Fig. 3.14d for (100) planes reveals the distorted core region which may again be due to the high Debye–Waller factor as seen in Table 3.11. The atomic number of Te is high which can be visualised from the dense distribution of charges in the electron density map. Since the atomic numbers of Sn and Te are higher than that of Mg, Ti, Fe and Zn, more diffuse distribution of charges are seen in the electron density maps of Sn and Te on the (100) plane.

The electron density map of Fe on (110) plane shows (Fig. 3.15c) highly concentrated charges at the core of the atoms, and the distribution of charges on places other than the atomic positions, due to the valence, free electrons. The electron density map of Mg, Ti, Zn, Fe on the (110) plane (Fig. 3.15a, d), show similar trends as observed in (100) plane. The electron density map of Sn and Te on the (110) plane (Fig. 3.15e, f), show less dense charge distribution compared to (100) plane.

The one-dimensional profiles of electron density constructed along [100], [110] and [111] directions for Mg, Ti, Fe, Zn, Sn and Te are shown in Fig. 3.16a–f, respectively. The positions of minimum electron densities and density values have been given in Table 3.14.

The electron densities at the saddle points along these three directions have been given in Table 3.13.

The mid-bond density in Ti is the largest value (Table 3.13) along [110] direction among the six metal systems studied. The Debye–Waller factor is less for Ti when compared with Mg, Fe, Zn, Sn and Te as shown in Table 3.11. Since the thermal vibration is less the bonding interaction is more significant in Ti. Hence, obviously, the bonding interaction is stronger in Ti as evidenced by the higher electron density along [110] direction. From Table 3.13, Zn has minimum mid-bond density of $0.00 \ 10^{-3}$ e/Å^3 at a distance of 0.4593 Å, among the metals studied along [110] and this may be due to the high Debye–Waller factor as observed in Table 3.11. Sn has less mid bond density $0.02 \ 10^{-3}$ e/Å^3 at a distance of 2.1796 Å, along [110] due to the high Debye–Waller factor as observed in Table 3.11. This substantiates the two-dimensional electron density map of tin (Fig. 3.15e) where sparse distribution of electron density is observed along [110] direction. From Tables 3.11 and 3.13, it can be observed that the metals with higher Debye–Waller factors have low mid-bond density along [110] direction which confirms the higher thermal vibration.

But along the [100] (nonbonding), Sn has higher electron density of $0.33 10^{-3}$ e/Å^3 at a distance of 1.6319 Å whose atomic number is higher than Mg, Ti, Fe, Zn. This can also be visualized from Fig. 3.14e which gives the MEM electron density distribution of Sn on the (100) plane.

Along [111] direction Fe has higher electron density of 0.52 e/Å^3 at a distance of 1.2424 Å which is also visualized in Fig. 3.15c which gives the MEM electron density distribution of Fe on the (110) plane.

Hence, from the qualitative and quantitative analysis of the MEM electron densities, the bonding in Mg, Ti, Fe, Zn, Sn and Te is predicted to be predominantly metallic and the interaction of charges along other non-bonding directions seems to be heavier due to the metallic nature.

3.3.7.2 PDF

In this study, the observed PDFs of Mg, Ti, Fe, Zn, Sn and Te were obtained from the raw intensity data using a software programme PDFgetX (Jeong et al. 2001) after performing corrections such as multiple scattering correction, polarisation correction, absorption correction, normalisation correction and Compton correction. A comparison between the observed and calculated PDF is carried out using the software package PDFFIT (Farrow et al. 2007). Figure 3.17a–f give the observed and calculated PDF profiles along with the difference between them for Mg, Ti, Fe, Zn, Sn and Te respectively.

The bond lengths deduced from the PDF analysis have been compared with those calculated using software programme Grenoble Thermal Ellipsoids Plot Programme (GRETEP) is a Windows interactive programme that displays the list

Table 3.14 Neighbour distances from PDF

Metal	Neighbour distance (from PDF analysis) (Å)			
	Nearest neighbour	Observed (r_1)	Calculated[a] (r_2)	Difference ($r_1 \sim r_2$)
Mg	I	3.12	3.21	0.09
	II	5.70	5.57	0.13
	III	8.24	8.46	0.22
Ti	I	2.80	2.95	0.15
	II	5.22	5.12	0.10
	III	7.36	7.63	0.27
Fe	I	2.80	2.85	0.05
	II	4.78	4.76	0.02
	III	6.44	6.25	0.19
Zn	I	2.62	2.67	0.05
	II	4.58	4.63	0.05
	III	5.44	5.64	0.20
Sn	I	2.38	2.19	0.18
	II	4.32	4.66	0.34
	III	6.38	6.21	0.17
Te	I	1.46	1.98	0.52
	II	3.32	3.96	0.64
	III	4.46	4.46	0.00

[a] GRETEP software

of calculated bonds, bond length between two atoms and the distances and angles around an atom) (GRETEP). The refined pair distribution function (the Fourier transform of, the reduced structure factor) given in Fig. 3.17a–f for Mg, Ti, Fe, Zn, Sn and Te show good matching of observed and calculated PDFs. The difference between the observed and calculated PDFs are also shown in the figures. The peaks in the PDF profile correspond to the nearest neighbour distances given in Table 3.14.

The first neighbour distance as observed from Table 3.15 decreases as the atomic number increases for all the metals. The difference between observed and calculated nearest neighbour distances are also given in the table. These differences are not too high considering the fact that we have used only data with limited Q values. There may be local undulations in the structure of these six systems that will lead to slightly higher neighbour differences which correspond to lattice repeat distances. The local disorder would have arisen due to several factors viz., X-ray diffraction instrumental error or fine impurity in the sample. This is reflected in the local structural analysis only like PDF.

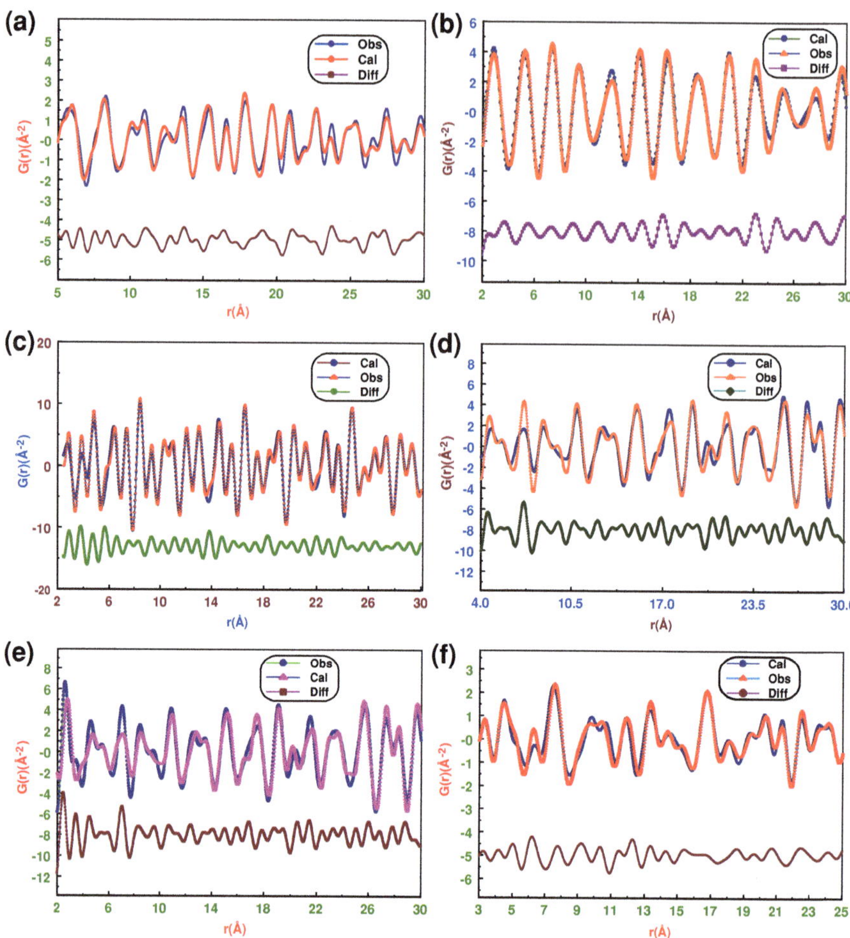

Fig. 3.17 **a** The observed and calculated PDFs of Mg with difference profile. **b** The observed and calculated PDF of Ti with difference profile. **c** The observed and calculated PDFs of Fe with difference profile. **d** The observed and calculated PDFs of Zn with difference profile. **e** The observed and calculated PDFs of Sn with difference profile. **f** The observed and calculated PDFs of Te with difference profile

3.3.8 Conclusion

A clear understanding of the structure of Mg, Ti, Fe, Zn, Sn and Te has been gained from this work, in terms of the cell parameters, thermal vibrations and electron density distribution in the bonding region using MEM and local structural analysis using PDF. The electron density, bonding and charge transfer studies analysed in this work will give fruitful information to researchers in the fields of physics, chemistry, materials science, metallurgy, etc. These properties can be properly used for the proper engineering of these technologically important materials.

3.4   CoAl and NiAl Metal Alloys

3.4.1   Introduction

Metal alloys are highly useful because they have specific properties and characteristics that are more attractive than those of pure, elemental metals. For example, some alloys possess high strength, some have low melting points, some are refractory with high melting temperatures, some are especially resistant to corrosion and some have desirable magnetic, thermal or electrical properties. These characteristics arise from both the internal and electronic structure of the alloys. Hence, a study on the electronic structure of alloys using the most versatile techniques like MEM (Collins 1982) and multipole method (Hansen and Coppens 1978) available today is worthwhile. If the tool available for the analysis yields highly precise information, then it is appropriate to apply it to precise data sets available as has been done in this work, thereby the methodology can also be tested. Hence, a study on the electronic structure of the metal alloys CoAl and NiAl has been carried out using accurate single crystal X-ray data sets reported earlier by Weiss (1966). Electron density distribution and bonding in different materials using MEM have been analysed for different materials e.g., GaAs (Saravanan et al. 2003), InP (Israel et al. 2004), Oxides (Israel et al. 2003), ionic systems (Israel et al. 2003). Research works have also been reported by Sakata and Sato (1990), Saka and Kato (1986), Yamamura et al. (1998), Ikeda et al. (1998), Yamamoto et al. (1996). In order to confirm the results and to elucidate the distribution of valence electrons, and the contraction/expansion of atomic shells, multipole analysis of the electron densities was carried out using the software package JANA 2000 (Petříček et al. 2000). Recently, the multipole analysis of the charge densities and bonding has been widely utilised to study the electronic structure of materials (Coppens and Volkov 2004; Poulsen et al. 2004; Frils et al. 2004; Pillet et al. 2004; Marabello et al. 2004). In this work, the multipole model proposed by Hansen and Coppens (1978) has been used for elucidating the electronic structure of CoAl and NiAl.

3.4.2   Summary of the Work

The precise electron density distribution and bonding in metal alloys CoAl and NiAl is characterised using MEM and multipole method. Reported X-ray single crystal data have been used for this purpose. Clear evidence of the metal bonding between the constituent atoms in these two systems has been obtained. The mid-bond electron densities in these systems are found to be 0.358 and 0.251 e/Å^3 respectively, for CoAl and NiAl in the MEM analysis. The two-dimensional maps and one-dimensional electron density profiles have been constructed and analysed. The thermal vibration of the individual atoms Co, Ni and Al has also been studied and reported. The contraction of atoms in CoAl and expansion of Ni and contraction of Al atom in NiAl is found from the multipole analysis, in line with the MEM electron density distribution.

3.4.3 Origin of the Data

Weiss (1966) has reported single crystal X-ray structure factor data of CoAl and NiAl collected using AgKα and CoKα radiations. Among these highly precise data sets, structure factor sets collected using AgKα were utilised in this work, because, more number of Bragg reflections is available using this radiation.

The two sets of data were subjected to preliminary data analysis and the structure of CoAl and NiAl was refined using the software GSAS (Larson and Von Dreele 2004) (General Structure Analysis System). Further, the same data were refined using a code written by Dr. R. Saravanan using an algorithm which uses the standard least-squares refinement. The individual thermal vibration parameters of atoms were refined both in GSAS (Larson and Von Dreele 2011) and also using the code by Dr. R. Saravanan (hereafter referred as RS refinements). Exact matching of the structure factors and structural parameters were realised in both the refinements. The refined structure factors were used for MEM analysis. Multipole refinements were carried out using the refinement software package JANA 2000 (Petříček et al. 2000) to refine the thermal parameters and the κ' parameter which gives information about the valence electronic status of the atomic shells.

3.4.4 Data Analysis

CoAl and NiAl belong to the CsCl structure in which there are 2 atoms/cell and the fractional atomic coordinates are 0, 0, 0, and ½, ½, ½. The cell constants are 2.8619 and 2.8864 Å respectively, for CoAl and NiAl, (Weiss 1966). Figure 3.18 shows the structure of NiAl.

It has been reported that no order–disorder transition occurs in these alloys and there is a possibility of squeezing of charge densities and directionality in the outer electron charge density. A total of 11 Bragg reflections have been reported (Weiss 1966) for both these systems. The data were found to be precise and the reported σ (F_{obs}) values are very low indicating the precision of the data. Using these data, first, the refinements were carried out with GSAS and the results of these refinements are given in Table 3.15 both for CoAl and NiAl. Then RS refinements were carried out, the results of which are also given in Table 3.15. The quantity

Table 3.15 The Debye–Waller factors in NiAl and CoAl in GSAS and RS Refinements

Refinement type	System			
	NiAl		CoAl	
GSAS	B_{Ni} (Å^2)	0.221(0.070)	B_{Co} (Å^2)	0.368(0.049)
	B_{Al} (Å)^2	0.421(0.144)	B_{Al} (Å^2)	0.591(0.088)
	R (%)	1.5	R (%)	1.1
	wR (%)	1.9	wR (%)	1.2
RS	B_{Ni} (Å^2)	0.279(0.073)	B_{Co} (Å^2)	0.353(0.082)
	B_{Al} (Å)^2	0.389(0.120)	B_{Al} (Å^2)	0.528(0.130)
	$B_{overall}$ (Å)^2	0.288(0.035)	$B_{overall}$ (Å)^2	0.382(0.042)
	R (%)	1.413	R (%)	1.134

B Debye–Waller factor

Table 3.16 Parameters of MEM refinement

Parameters	NiAl	CoAl
Prior electron density (e/Å^3)	1.705	1.706
Resolution (Å/Pixel)	0.405	0.405
Λ	0.24	0.24
C	1.0	1.0
R (%)	0.88	0.92
wR (%)	0.80	0.79
Number of cycles	223	188

λ Lagrangean parameter; C convergence criterion

Fig. 3.18 Unit cell of NiAl

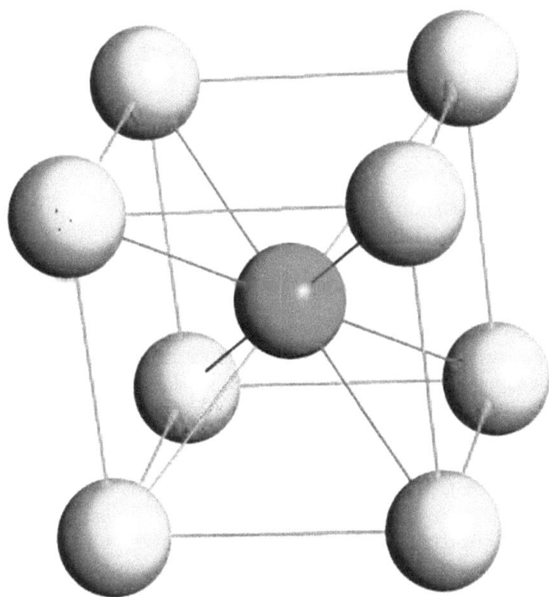

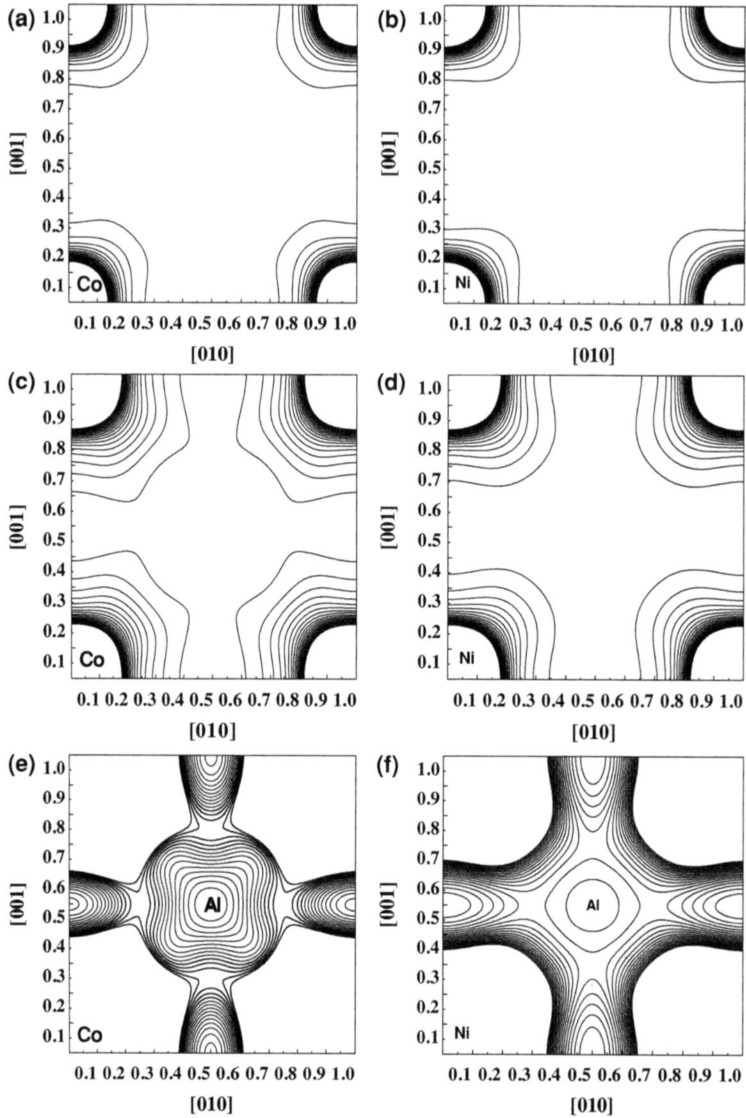

Fig. 3.19 **a** MEM high electron density distribution of CoAl on the (100) plane. Contour Range is between 0.0 and 15.0 e/Å^3. Contour interval is 0.75 e/Å^3. **b** MEM high electron density distribution of NiAl on the (100) plane. Contour range is between 0.0 and 15.0 e/Å^3. Contour interval is 0.75 e/Å^3. **c** MEM low electron density distribution of CoAl on the (100) plane. Contour range is between 0.0 and 2.5 e/Å^3. Contour interval is 0.125 e/Å^3. **d** MEM low electron density distribution of NiAl on the (100) plane. Contour range is between 0.0 and 2.5 e/Å^3. Contour interval is 0.125 e/Å^3. **e** MEM electron density distribution of CoAl on the (100) plane showing the extent of electron clouds of Al. Contour range is between 0.0 and 0.2 e/Å^3. Contour interval is 0.008 e/Å^3. **f** MEM electron density distribution of NiAl on the (100) plane showing the extent of electron clouds of Al. Contour range is between 0.0 and 0.2 e/Å^3. Contour interval is 0.008 e/Å^3

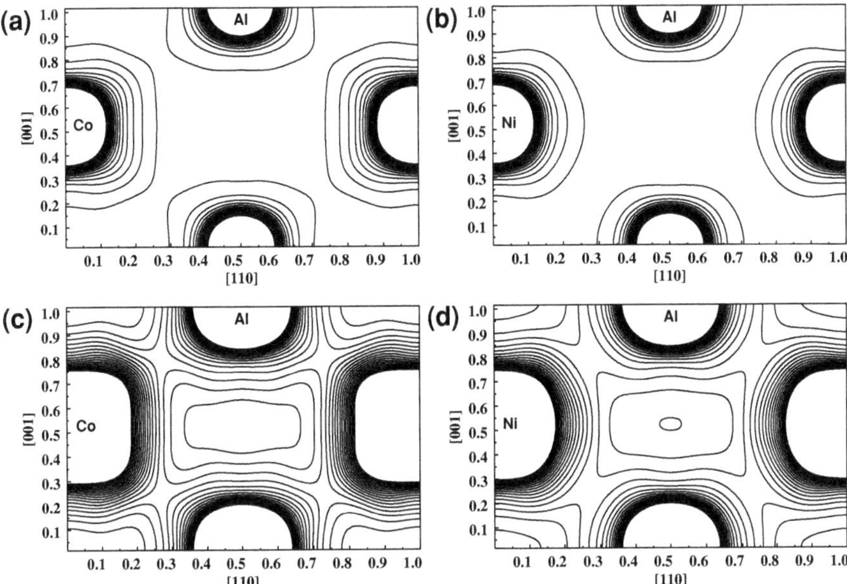

Fig. 3.20 a MEM high electron density distribution of CoAl on the $(1\bar{1}0)$ Plane. Contour range: 0.0–6.6 e/Å^3. Contour interval is 0.33 e/Å^3. The plane has been shifted half the unit cell along the z direction. **b** MEM high electron density distribution of NiAl on the $(1\bar{1}0)$ Plane. Contour range: 0.0–6.6 e/Å^3. Contour interval is 0.33 e/Å^3. The plane has been shifted half the unit cell along the z direction. **c** MEM low electron density distribution of CoAl on the $(1\bar{1}0)$ Plane. Contour range: 0.0–1.2 e/Å^3. Contour interval is 0.06 e/Å^3. The plane has been shifted half the unit cell along the z direction. **d** MEM low electron density distribution of NiAl on the $(1\bar{1}0)$ Plane. Contour range: 0.0–1.2 e/Å^3. Contour interval is 0.06 e/Å^3. The plane has been shifted half the unit cell along the z direction

minimised in RS refinements was $D = \sum_{hkl} W_{hkl}(|F_{obs}| - |kF_{cal}|)^2$, where w_{hkl} is the weight assigned on each observation, F_{obs} is the observed structure factor and F_{cal} is the calculated structure factor. The estimated standard deviation in any parameter is given by $\sigma_p = \left[b_{ii} \sum_{r=1}^{w} \frac{(\Delta F_r)^2}{m-n} \right]$, where b_{ii} is the diagonal element of the inverse matrix, w_r is the weight of the rth factor, where ΔF_r is the difference between F_{obs} and F_{cal}, m is the number of observations and n is the number of parameters. The reliability index given by $R = \dfrac{\sum ||F_{obs}| - |F_{cal}||}{\sum |F_{obs}|}$ of these refinements is also given in Table 3.15. The average thermal vibration of the overall system is studied and reported in Table 3.15. In RS refinements, the analytical coefficients of the five-parameter model by Waasmaier and Kirfel (1995) were used for the calculation of the atomic scattering factors.

MEM studies on the electron density distribution of these two systems were carried out by dividing both the unit cells into $64 \times 64 \times 64$ pixels. The initial electron densities were fixed as (F_{000}/a_o^3), where F_{000} is the total number of

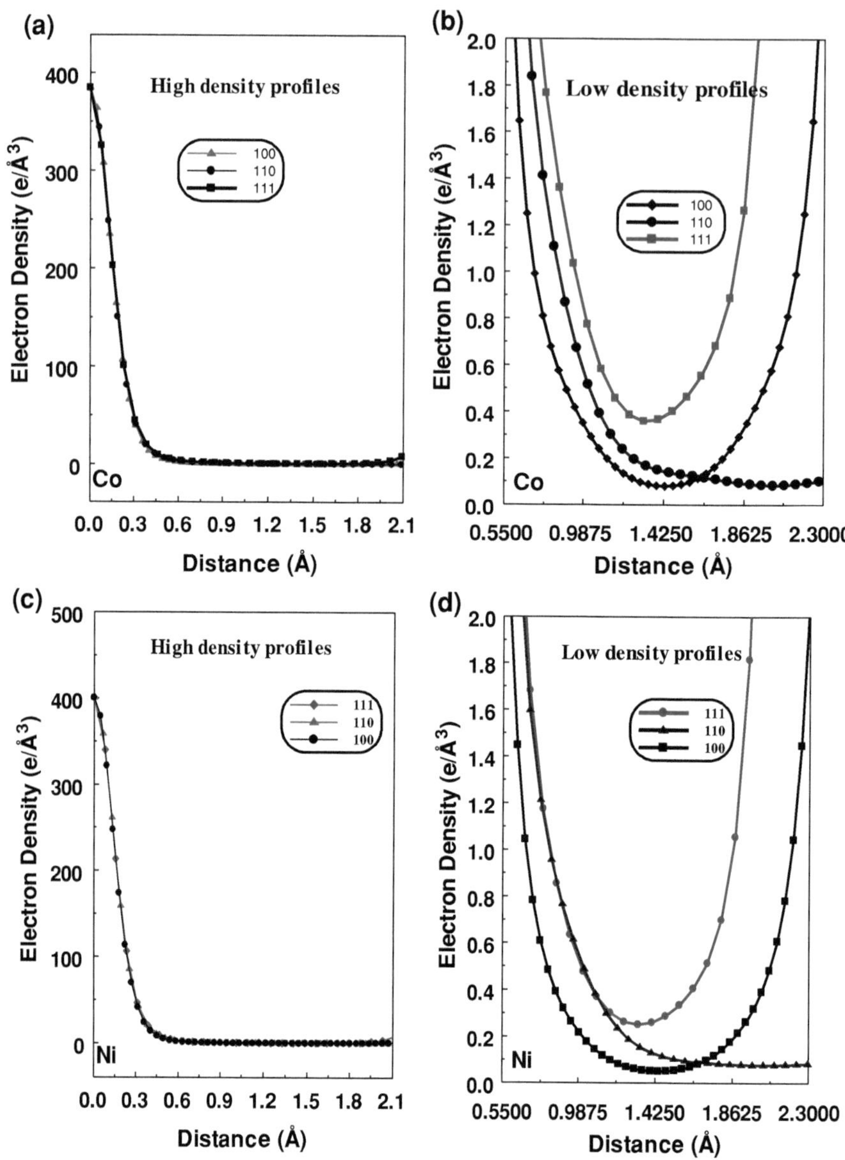

Fig. 3.21 **a** One-dimensional high electron density distribution of CoAl along the [100], [110] and [111] directions of the unit cell. **b** One-dimensional low electron density distribution of CoAl along the [100], [110] and [111] directions of the unit cell. **c** One-dimensional high electron density distribution of NiAl along the [100], [110] and [111] directions of the unit cell. **d** One-dimensional low electron density distribution of NiAl along the [100], [110] and [111] directions of the unit cell

Table 3.17 Profile electron densities along the three directions of the unit cells of CoAl and NiAl

Direction system	[100]		[110]		[111]	
	Position (Å)	ED (e/Å^3)	Position (Å)	ED (e/Å^3)	Position (Å)	ED (e/Å^3)
CoAl	1.431	0.079	2.024	0.083	1.317	0.358
NiAl	1.443	0.050	2.041	0.078	1.328	0.251

ED Electron density

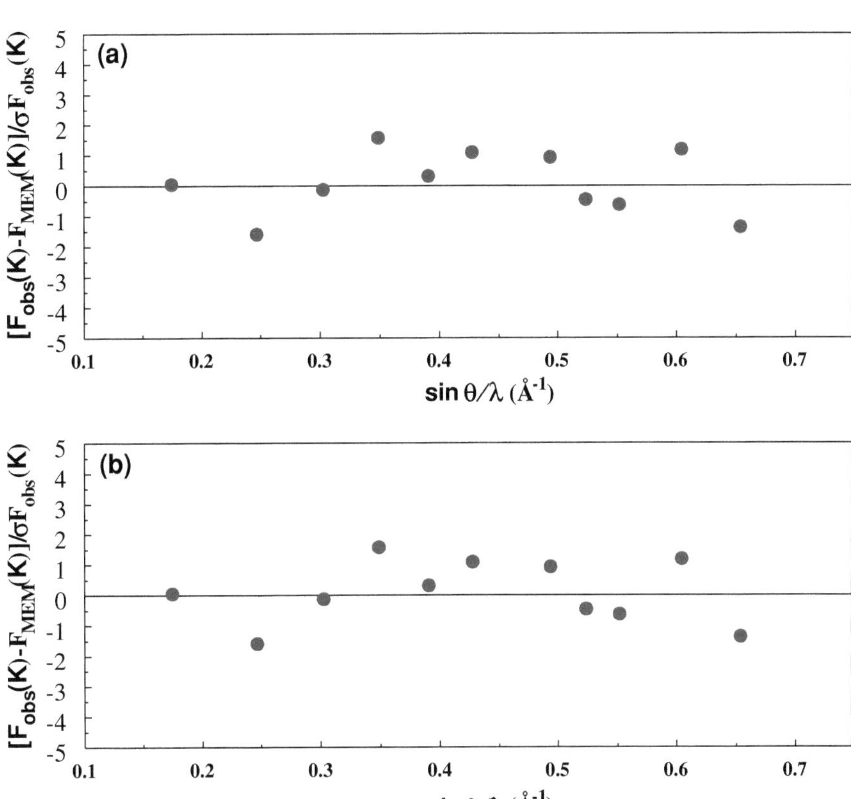

Fig. 3.22 a The distribution of errors in the reciprocal space of CoAl. **b** The distribution of errors in the reciprocal space of NiAl

electrons in the respective unit cells and a_o is the cell dimension. The resolution of the MEM maps turns out to be 0.045 Å/pixel along the three directions of the unit cells. The Lagrange parameter λ was chosen so that 188 and 223 cycles of refinement were required for CoAl and NiAl respectively, to achieve the convergence criterion C = 1; C being,

Table 3.18 Parameters from Multipole refinement

Parameter	NiAl		CoAl	
	Ni	Al	Co	Al
Pc	18.0	10.0	18.0	10.0
Pv	10.0	3.0	9.0	3.0
κ'	0.984(047)	1.258(087)	0.953(039)	0.920(096)
B (Å^2)	0.297(076)	0.337(128)	0.265(040)	0.472(065)
R (%)	1.08		0.91	
wR (%)	1.10		0.97	
GoF	1.71		1.55	

B Debye–Waller factor, Pc population coefficients in the core region, Pv population coefficients in the valence region; κ' and κ'' variable parameters

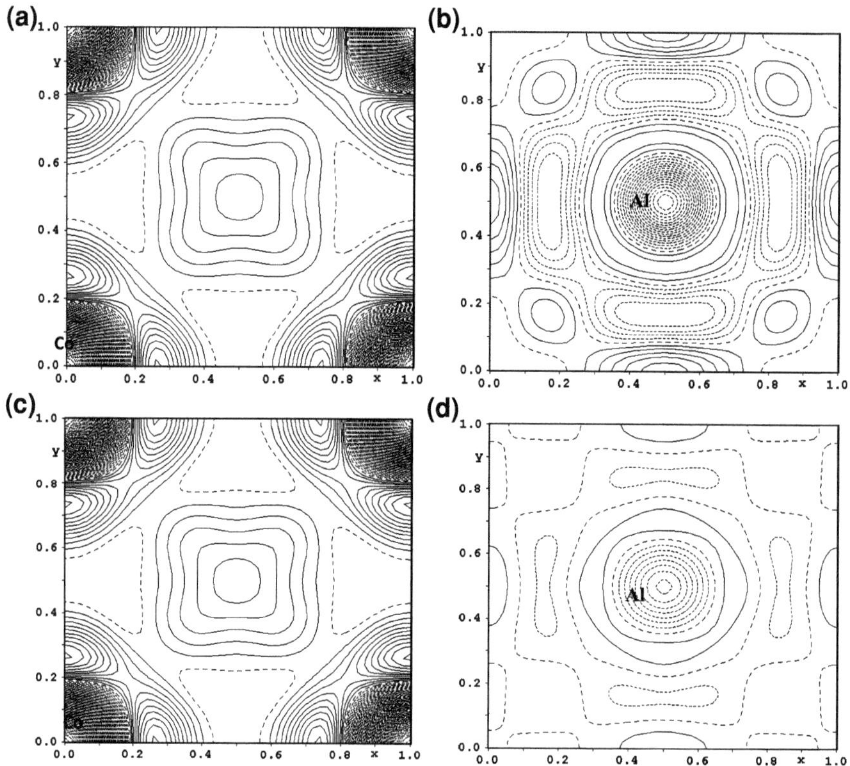

Fig. 3.23 a Static Multipole Deformation map of CoAl on (100) plane. *Dotted lines* indicate negative electron densities. **b** Static Multipole Deformation map of CoAl on (½00) plane. *Dotted lines* indicate negative electron densities. **c** Dynamic Multipole Deformation map of CoAl on (100) plane. *Dotted lines* indicate negative electron densities. **d** Dynamic Multipole Deformation of CoAl on (½00) plane. *Dotted lines* indicate negative electron densities

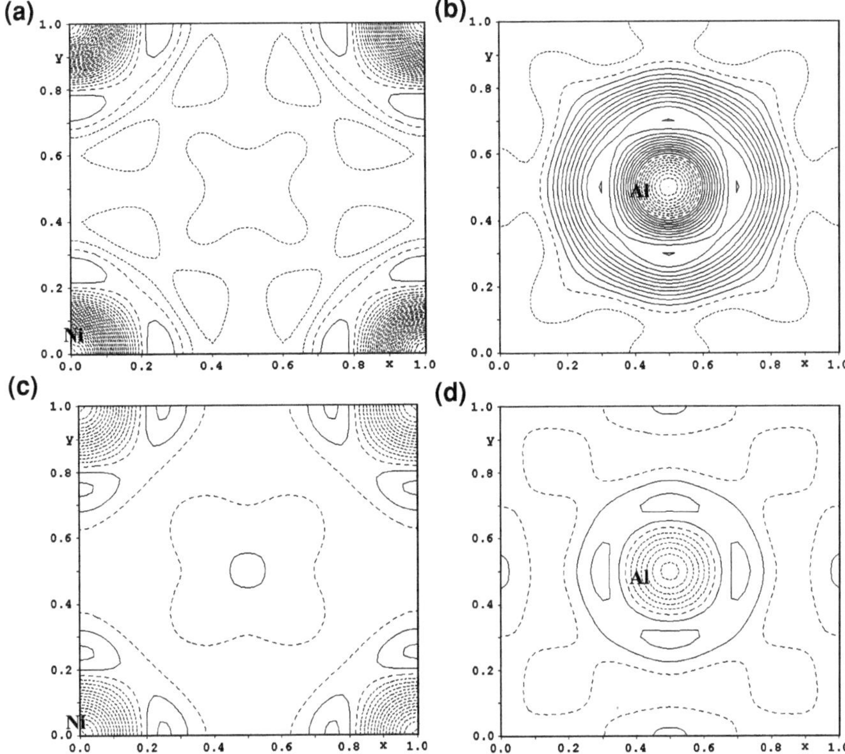

Fig. 3.24 a Static Multipole Deformation map of NiAl on (100) plane. *Dotted lines* indicate negative electron densities. **b** Static Multipole Deformation map of NiAl on (½00) plane. *Dotted lines* indicate negative electron densities. **c** Dynamic Multipole Deformation map of NiAl on (100) plane. *Dotted lines* indicate negative electron densities. **d** Dynamic Multipole Deformation map of NiAl on (½00) plane. *Dotted lines* indicate negative electron densities

$$C = \frac{1}{N} \sum \left| \frac{F_{\text{cal}}(k) - F_{\text{obs}}(k)^2}{\sigma(F_{\text{obs}}(k))} \right| \tag{3.5}$$

where N is the number of Bragg reflections, $\sigma(K)$ is the standard deviation in $F_{obs}(K)$ and V is the volume of the unit cell. Other relevant parameters of MEM refinements are given in Table 3.16.

The electron density distribution of CoAl and NiAl has been mapped using the MEM electron density values obtained through refinements. Figure 3.19a, b show the MEM high electron density distribution of CoAl and NiAl on the (100) plane of the unit cell.

Figure 3.19c, d show the MEM low electron density distribution of CoAl and NiAl on the (100) plane.

Figure 3.19e, f show the extent to which the aluminium atom is distributed as evidenced by the electron clouds on the (100) plane.

Figure 3.20a, b show the MEM high electron density distribution on the ($1\bar{1}0$) plane of CoAl and NiAl respectively, wherein the shifted atoms (half the unit cell along the Z direction) have been shown.

Figure 3.20c, d show the MEM low electron density distribution on the ($1\bar{1}0$) plane of CoAl and NiAl.

Figure 3.21a, b show the one-dimensional variation of the high and low electron densities along the three directions ([100], [110] and [111]) of the unit cell of CoAl and Figs. 3.21c, d represent the one-dimensional variation of high and low electron densities of NiAl.

The electron densities at the saddle points along these three directions are given in Table 3.17.

Based on these two-dimensional MEM maps and one-dimensional profiles, a discussion is formulated and given in the following section.

The errors in the observations have also been studied in this work and represented in Fig. 3.22a, b, which show the distribution of errors in the reciprocal space CoAl and NiAl respectively.

The results of the multipole refinements done using JANA 2000 are presented in Table 3.18.

Hansen and Coppens (1978) proposed a modified electron density model with the option that allows the refinement of population parameters at various orbital levels where the atomic density is described as a series expansion in real spherical harmonic functions through fourth order Y_{lm}. According to this model, the charge density in a crystal is written as the superposition of harmonically vibrating aspherical atomic density distribution convolving with the Gaussian thermal displacement distribution as

$$\rho(r) = \sum_{k}^{atom} \rho_k\left(\vec{r} - \vec{r}_k - \vec{u}\right) \otimes t_k\left(\vec{u}\right), \tag{3.6}$$

where $t_k(\vec{u})$ is the Gaussian term and $\otimes$ indicates a convolution. The atomic charge density is then defined as

$$\rho\left(\vec{r}\right) = P_c \rho_{core}\left(\vec{r}\right) + P_v k'^3 \rho_{valence}(k'r) + \sum_{l=0}^{4} k''^3 R_j(k''.) \sum_{=-l}^{l} P_{lm} Y_{lm}\left(\frac{\vec{r}}{r}\right) \tag{3.7}$$

where P_c, P_v and P_{lm} are population coefficients. Canonical Hartree–Fock atomic orbitals of the free atoms normalised to one electron can be used for the construction of ρ_{core} and $\rho_{valence}$ but the valence function is allowed to expand and contract by the adjustment of the variable parameters $k\prime$ and $k\prime\prime$. The effect of the temperature can be distinguished from the convolution and the deconvolution of thermal contribution to the charge density as dynamic and static multipole deformation maps. The deformation density in these maps are characterised by

$$\Delta\rho_{\text{multipole\_deformation}}\left(\overrightarrow{r}\right) = \frac{1}{v}\sum\left[F\left(\overrightarrow{h}\right)_{\text{multipole\_deformation}}\right.$$

$$\left. - F\left(\overrightarrow{h}\right)_{\text{spherical\_atom}}\right]\exp\left[-2\pi i\left(\overrightarrow{h},\overrightarrow{r}\right)\right] \quad (3.8)$$

where $F_{\text{multipole\_deformation}}$ is the Fourier transform of the multipole charge density with or without the convolution of thermal contribution, where the Fourier components are terminated at the experiment resolution. The SMD and DMD maps are presented in Fig. 3.23a, b, c, d for CoAl and in Fig. 3.24a, b, c, d for NiAl. That is, Fig. 3.23a, b represent the SMD map of CoAl on the (100) and (½00) planes; Fig. 3.24a, b represent the SMD map of NiAl on the (100) and (½00) planes. Other maps are DMD maps of these two systems. The static deformation maps show the electron distribution without the contribution from thermal vibration of the atoms.

3.4.5 Results and Discussion

The structural refinement gives precise individual Debye–Waller factors as indicated in Table 3.15. The Al atom has a large thermal vibration in both CoAl and NiAl compared to the constituent atoms Co and Ni. The B_{Al} is larger in CoAl than in NiAl, indicating slightly larger binding force in NiAl due to more number of electrons. The reliability indices in these refinements are as low as 1.1% in the case of CoAl. The RS refinements give similar results for the thermal vibrations. The reliability indices are also low compared to GSAS refinements in the case of CoAl, whereas in NiAl, the R values are comparable in both the refinements. The difference in B_{Ni} between these two refinements is 0.058 Å^2, larger in the case of RS refinement. But B_{Co} is 0.015 Å^2 smaller in RS refinement compared to that of GSAS refinement. The difference in B_{Al} between NiAl and CoAl is 0.17 Å^2 in GSAS refinement and 0.139 Å^2 in RS refinement. The fact that the atom with smaller atomic number vibrates more has been realised in both the refinements and in both CoAl and NiAl ($Z_{Al} < Z_{Co} < Z_{Ni}$).

The MEM refinements were carried out with a prior electron density of 1.706 and 1.705 e/Å^3 respectively, for CoAl and NiAl. The Lagrange parameter used was the same for both the cases. The reliability indices from MEM refinements are given in Table 3.16, which shows very low R values, being less than 1%.

Figure 3.19a, b shows highly spherical core regions of Co and Ni atoms respectively, on the (100) plane. The contour range and interval for these two maps were set as the same in order to compare the electron densities. The same criterion was followed for appropriate low and high-density maps of CoAl and NiAl. In order to visualise the low electron density/bonding regions, the maps were re-drawn with a contour range of 0.0–2.5 e/Å^3, as shown in Fig. 3.19c, d. Figure 3.19c shows that the valence region of Co is clearly not spherical and the four corner atoms on (100) plane show concave outer charges. It appears that the eight Co atoms give way for the thermal vibration of the Al atom located at the body centre. It is noted from Table 3.15 that B_{Al} in CoAl is higher than B_{Co}. In the case of NiAl (Fig. 3.19d), no such concave outer charges are seen, since the thermal vibration of both Ni and Al atoms are lower compared to Co and Al in CoAl. Figure 3.19e, f show very low electron densities on (100) plane of CoAl and NiAl. The body-centred Al charges touching the (100) plane of the unit cell is visible in Fig. 3.19e. The valence charges of both the atoms (Co and Al) do not repel each other. In fact, charge clouds of both the atoms touch each other, but with no resemblance of an attraction. This should be due to the characteristics of metal bonding. Moreover, in multipole analysis, the expansion/contraction parameter κ' is estimated to be 0.953 and 0.92 for Co and Al atoms, which means expansion of both the atoms. The expansion is more in Al atom. Hence, the above said trends in the low-density regions are visible as seen from Fig. 3.19e. In NiAl (Fig. 3.19f), the charge accumulation is less dense due to reduced thermal vibration of both the atoms in NiAl. Again from multipole refinement, the κ' parameter is found to be 0.984 and 1.258 for Ni and Al atoms respectively. This clearly shows that Ni atom is expanded slightly (not as in Co), but the Al atom is contracted to a large extent ($\kappa' = 1.258$). These facts are clearly visualized in the MEM maps. Thus, the results from one analysis (MEM) could be correlated to the other (multipole). Figure 3.20a, b show the electron density distribution of CoAl and NiAl in the high-density regions of the $(1\bar{1}0)$ plane. The core regions of both the atoms are seen to be spherical both in CoAl and NiAl. Again, no repulsion of charges is seen in these systems. The outer regions of both Co and Al atoms are more smeared in CoAl compared to that of Ni and Al atoms in NiAl, a fact due to the increased thermal vibrations in CoAl. Figure 3.20c, d show the low electron density distribution of CoAl and NiAl on the $(1\bar{1}0)$ plane. Charges are seen around the periphery of Co and Al atoms in CoAl (Fig. 3.20c) and Ni and Al atoms in NiAl (Fig. 3.20d). The bonding charges show a metallic nature in both the systems.

Figure 3.21a–d show the one-dimensional electron density distribution along the [100], [110] and [111] directions of the CoAl and NiAl unit cells respectively. Both high- and low-density variation of charges is shown in these figures. The high-density peaks in Fig. 3.21a, c show the peak densities along these three directions which are larger in NiAl. The low-density maps show that the density minima do not occur at the same points in both the cases. Moreover, the distribution of charges along the three directions does not overlap. These facts suggest that the outer regions are not perfectly spherical. The profile along [100] direction is perfectly symmetrical due to the distribution of charges from like atoms. The distribution along [111] direction shows the bonding nature between unlike atoms. The previous discussion clears that there is attraction/repulsion like interaction between Co and Al and Ni and Al. Yet, the mid-bond charges are considerably higher as seen from Table 3.17. CoAl has a mid-bond density value of 0.358 e/Å^3

and NiAl has a value of about 0.251 e/$\mathring{A}^3$. If the systems are ionic one cannot expect higher mid-bond values as in the present cases. If the systems are covalent, one can expect values higher than the present values due to overlapping of charges. Hence, from the qualitative and quantitative analysis of the MEM electron densities, the bonding in CoAl and NiAl is predicted to be predominantly metallic. The strength of the electron density is higher in the case of CoAl, which has lower number of electrons in the unit cell compared to NiAl. This is due to the relatively larger thermal vibration of aluminium and smearing of charges in CoAl. The increase in the mid-bond density is about 0.107 e/$\mathring{A}^3$ consistent with the increased thermal vibration of atoms, particularly aluminium, in CoAl. Hence, it is important to study precisely the thermal vibration of atoms in electron density analyses, particularly in comparing the bonding densities in systems with similar constituent atoms.

Figure 3.22a, b shows the distribution of errors in the observations in the reciprocal space of CoAl and NiAl respectively. The error in the form of $[F_{obs}(K) - F_{MEM}(K)]/\sigma(F_{obs}(K))$, is distributed uniformly around ± 2 indicating that there is no bias in the experimental data or the calculations and justifies the use of these experimental data for the present analysis.

The B values from the multipole refinement are in close agreement with those obtained from other refinements. Although all the B values are comparable, those from the multipole analysis give accurate interpretation of the electron density distribution in both the systems.

The SMD and DMD maps obtained from multipole refinements show that (Figs. 3.23, 3.24) more pronounced deformation in density is at the valence regions than at the core regions for both the systems. Figure 3.23a (SMD—CoAl—100 plane) shows features similar to those seen in Fig. 3.19e for CoAl. The expansion of Al atom and the concave nature of charge clouds of Co atom are visible in this map. Figure 3.24a (NiAl—SMD 100 plane) does not show any concave nature/repulsive interaction, because of the heavy contraction of Al atom. The map in Fig. 3.23b (SMD—CoAl—½ 0 0 plane) shows the expansion of both the atoms and Fig. 3.24b (SMD—NiAl—½00 plane) shows nothing but Al atom because of its contracted nature. Further, the expansion of Ni atom ($\kappa = 0.984$) is not large compared to Co (0.953) atom. Hence, only the distribution of Al atom is seen in this map. The DMD maps show trends in line with the expansion/contraction of atoms in the two systems. Comparing Fig. 3.23b (SMD of CoAl) and Fig. 3.23d (DMD of CoAl), it is seen that the SMD map shows a number of positive contours at the edge centres and at the centre of (100) plane. These are the residual densities due to the thermal vibration of aluminium atom. The core of aluminium atom in the Fig. 3.23b is full of negative charges. The DMD map shows (Fig. 3.23d) the uncompensated thermal vibration of the valence regions that show up as the positive contours starting from the core. This suggests slight expansion of the aluminium atom.

Figure 3.24b (SMD of NiAl) and 3.24d (DMD of NiAl) can be compared in a similar manner. Figure 3.24b shows positive electron densities not only at the core but also at far-valence regions, even after the deconvolution process, suggesting shrinkage of the aluminium atom in NiAl. The DMD (Fig. 3.24d) shows a well compensated core region (without negative charge densities) but uncompensated positive islands of valence region confirming the contraction of aluminium atom in NiAl.

3.4.6 Conclusion

A precise analysis on the thermal vibrations and electron density distribution in CoAl and NiAl has been carried out using reported X-ray data. The metal bonding in these two systems is substantiated using qualitative and quantitative analysis of MEM and multipole electron densities. CoAl is found to have more mid-bond density than NiAl due to the increased thermal vibration and hence charge delocalization. The cores of the atoms are found to be perfectly spherical and slight deviation is observed in the valence electron distribution. More studies are needed on more alloy systems in terms not only of electron densities but also in terms of electron densities coupled with thermal vibrations of the atoms involved.

3.5 Nickel Chromium ($Ni_{80}Cr_{20}$)

3.5.1 Introduction

Alloying one metal with other metals or non-metals often enhances its properties. The physical properties of an alloy may not differ greatly from those of its elements, but engineering properties may be substantially different from those of the constituent materials.

Whether in the form of element or alloy with other metals, nickel materials have made significant contributions to present-day society. The favourable physical and mechanical properties as well as corrosion resistance characteristics of nickel and its alloys provide aids to the solution of many industrial problems (Teeple 1954). Alloying with nickel increases the impact strength and ductility of sintered materials without adversely affecting their strength (Enríquez and Mathew 2003).

Nickel alloys have good corrosion resistance and heat resistance, the standard alloy being used for electrical-resistance for heaters and electrical appliances is Nickel–Chromium (Murray 1997). Chromium–nickel steel alloys have good mechanical properties (Radomysel''skii and Kholodnyi 1975). Conductive nichrome probe tips have been developed as functionalized nanotools for nano-scale electrical testing and nano-welding, and are advantageous in competition with other scanning probe tips because of their unique combination of high resistance to oxidation, high hardness and relatively high resistivity (Peng et al. 2009). Ni$_{80}$Cr$_{20}$ is a commonly used material for application in thin-film strain gauges (Kazi et al. 2003, 2006]. NiCr coatings on stainless steel substrates are used as a high velocity oxy-fuel technique for corrosion applications (Ak et al. 2003). NiCr cermet coatings are commonly utilised in the industry as a means to provide wear resistance in high temperature (Wirojanupatump et al. 2001). NiCr-coated super alloys show better oxidation resistance due to the formation of a compact and adhesive thin Cr$_2$O$_3$ scale on the surface of the coating during oxidation (Kamal et al. 2009).

For practical use in different applications, alloys need to be shaped at elevated temperatures according to the requirements. It is therefore, essential to study the changes in the microstructure after heat treatment of materials. One of the popular heat treatment processes is annealing (Al-Quran and Al-Itawi 2010). Annealing induces ductility, softens material, relieves internal stresses, refines the structure by making it homogeneous and improves cold working properties. The resistance characteristics of NiCr improve in the double annealing process (Jeng et al. 1990). Annealing processes appear to be responsible for an increase in passivity of Ni/Cr alloys toward oxidation (Kim et al. 2005).

Ball milling may be referred to as the breaking down of relatively coarse materials to the ultimate fineness. At present, it has evolved as an important method for the preparation of either materials with enhanced physical and mechanical properties or indeed, new phases or new engineering materials. The morphology of the powders is modified when they are subjected to collisions. The initial ball-powder-ball collision causes ductile metal powders to flatten into flake-like structures (Eskandarany 2001). The high-energy ball milling technique is an effective method for producing powders for metal matrix composites (Zwick and Lugscheider 2004). There exist methods to control the microstructure of pellet and among others, mechanical milling of powder is simple and easy to control. With this method, the characteristics of powder and the resultant compaction behaviour and sinterability are properly controlled (Na et al. 2002).

3.5.2 Summary of the Work

The alloy Ni$_{80}$Cr$_{20}$ was annealed and ball-milled to study the effect of thermal and mechanical treatments on the local structure and on the electron density distribution. The electron density between the atoms was studied by MEM and the local structure using PDF. The electron density is found to be high for ball-milled sample along the bonding direction. The particle sizes of the differently treated samples were realised by SEM and through XRD. Clear evidence of the effect of ball milling is observed on the local structure and electron densities.

In this work, the purchased alloy $Ni_{80}Cr_{20}$ was subjected to different types of mechanical and thermal treatments viz., ball milling and annealing. The three different processes are as follows:

i.   annealing the sample (hereafter sample A)
ii.   ball milling (hereafter sample BM)
iii. ball milling and annealing (hereafter sample BMA).

The charge density and local structure variations of these three samples have been analysed in this work. This analysis would reveal the properties of $Ni_{80}Cr_{20}$ suitable for its wide applications. The electronic structure analysis has been carried out by the most versatile technique, MEM (Gull and Daniel 1978) and local and average structure properties have been analysed using PDF (Petkov et al. 2002; Xiangyun et al. 2005).

3.5.3 Experimental

3.5.3.1 XRD and SEM

Powder sample of $Ni_{80}Cr_{20}$ (99.9% purity) was purchased from Alfa Aesar, Johnson's Mathew Company. To anneal the sample, 20 mg of the sample was placed in a small silica crucible and was annealed at 1023 K for 3 h in a programmable furnace (sample A). To prepare the ball-milled sample, the alloy was milled at room temperature in a low energy laboratory ball mill for 40 h with a milling speed of 170 rpm, using stainless steel balls (sample BM). Further, a part of the ball-milled powder was annealed at 1023 K for 3 h (sample BMA). The low energy laboratory ball mill used to mill the sample and the furnace in which the samples were annealed are shown in Figs. 3.25a, b, respectively.

To analyse the bonding charge densities (structural behaviour) of $Ni_{80}Cr_{20}$ under various experimental conditions, powder X-ray diffraction data were collected individually for sample A, sample BM, sample BMA. X-ray powder diffraction measurements were performed using X-PERT PRO (Philips, Netherlands) at the National Institute of Interdisciplinary Science and Technology (NIIST), Trivandrum, India using $CuK\alpha_1$ radiation with a 2θ range from 10 to 120° and 0.05° step size.

Particle size is an important parameter that influences the physical properties of materials. A particle may be made of several different crystallites. A crystallite size is the minimum part of the material that diffracts coherently. Several techniques could be used for the investigation of the particle size. X-ray diffraction (XRD) and Scanning electron microscopy (SEM) techniques are used in this work to obtain the particle size of the samples. From XRD data the particle size was obtained as the particle size is related to the diffraction peak

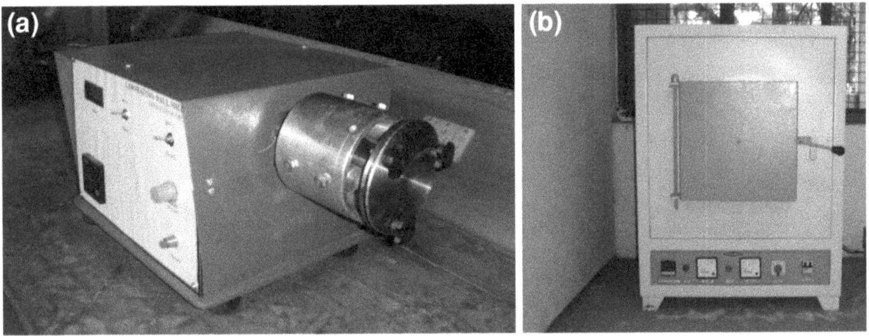

Fig. 3.25 a Laboratory ball mill. **b** Programmable high temperature furnace

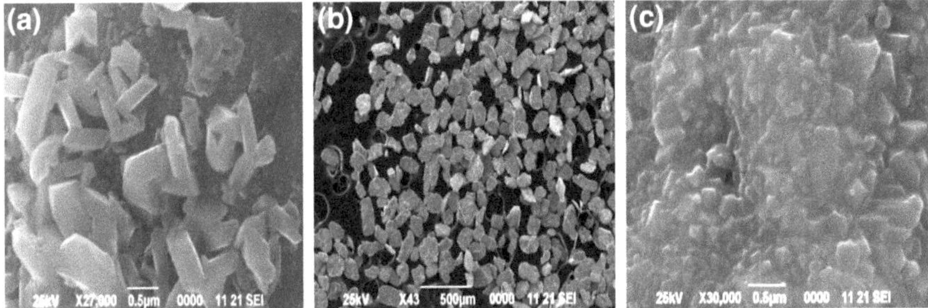

Fig. 3.26 a SEM picture of sample A. **b** SEM picture of sample BM. **c** SEM picture of sample BMA

Table 3.19 Particle size of the three samples

Sample	[a]Particle size (N_x) from XRD (nm)	Particle size (N_s) from (SEM) (nm)	Average number of coherent domains in each particle (N_s/N_x)
A	25.65	78.75	3.07
BM	11.45	38.76	3.39
BMA	19.80	59.35	3.00
			mean = 3.15

[a] coherently diffracting domains

broadening. The software written by Dr.R.Saravanan (GRAIN software) was used to estimate approximate particle sizes. The particle size is analysed using full width at half-maximum of the powder XRD peaks. The particle morphology was examined by SEM. The results of SEM analysis are given in Fig. 3.26a–c.

SEM pictures from Figs. 3.26a–c, show that the particle sizes are in nanoscale and sample A is elongated with particle size of 212.12 nm (Fig. 3.26a), sample BM is flattened with particle size of 38.76 nm (Fig. 3.26b) and sample

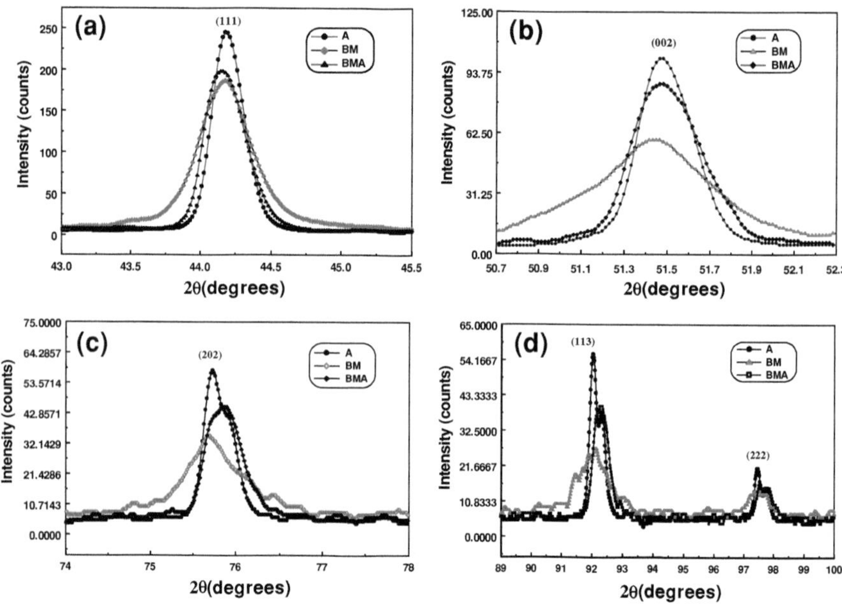

Fig. 3.27 a Comparison of (111) XRD peaks of the samples A, BM, BMA respectively.
b Comparison of (002) XRD peaks of the samples A, BM, BMA respectively. c Comparison of
(202) XRD peaks of the samples A, BM, BMA respectively. d Comparison of (113, 222) XRD
peaks of the samples A, BM, BMA respectively

Table 3.20 Full width at half-maximum of different miller planes

Sample	(111)	(002)	(202)	(113)	(222)
A	0.265	0.310	0.395	0.533	0.587
BM	0.440	0.707	0.892	1.477	1.912
BMA	0.347	0.396	0.527	0.642	0.801

BMA is with uniform morphology with particle size of 59.35 nm (Fig. 3.26c).
The obtained particle sizes from XRD and SEM are given in Table 3.19.

The sizes of the samples measured by XRD line broadening are about one-third
of the particle sizes calculated from SEM. The individual particles of the samples
may have been composed of coherently diffracting domains rather than each being
a single coherently diffracting crystal.

From Fig. 3.27a–d, it is observed that the intensity of samples decrease in the
order of A, BMA, BM respectively.

The broadening of the peaks is also observed in the case of sample BM. The
full width at half-maximum (FWHM) as observed in Table 3.20 indicates that
peaks are broader for sample BM compared to samples A and BMA. The
reduction in particle size due to ball milling as observed from SEM pictures is
substantiated.

3.5.4 Results and Discussion

3.5.4.1 Powder XRD Analysis

The structural parameters of the samples (A, BM, BMA) of Ni$_{80}$Cr$_{20}$ which have FCC structure with four atoms/unit cell in the space group Fm-3 m were refined with the well-known Rietveld (1969) powder profile fitting methodology using the software package JANA 2006 (Petříček et al. 2006). The fitted profiles and the positions of Bragg peaks for the three samples have been shown in Fig. 3.28a–c.

In these figures, the dots represent the observed powder patterns and the continuous lines represent the calculated powder profiles of the respective samples. The small vertical lines indicate Bragg positions. The difference between the observed profile and fitted calculated profile is shown at the bottom of each figure. The refined profiles show good matching between the observed and calculated profiles for all the three systems. The refined structure factors of the samples given in Table 3.21a–c shows very good agreement of the observed and calculated structure factors with very low statistical errors. The composition of the sample was also refined and was determined as Ni$_{80.1}$Cr$_{19.9}$ for the annealed sample (sample A).

The refined structural parameters from the software JANA 2006 are given in Table 3.22. The refinement indices are low which show the accuracy of the refinement (Rp = 8.17, 8.35, 10.16% respectively for samples A, BM, BMA). The cell parameters and thermal vibration parameters are observed to reduce in the order of sample BM, A, BMA respectively. Ball milling and subsequent annealing appears to have reduced the cell constant.

3.5.4.2 Charge Density of Ni$_{80}$Cr$_{20}$ Using MEM

The refined structure factors of the three samples were used for the MEM analysis. The software package PRIMA (Ruben and Fujio 2004) was used for the numerical MEM computations. The two- and three-dimensional representation of the electron densities were picturised using the software VESTA (Momma and Izumi 2006). The MEM refinements were carried out by dividing the unit cell into $64 \times 64 \times 64$ pixels.

The Lagrange parameter is suitably chosen so that the convergence criterion C = 1 is reached after minimum number of iterations. The MEM parameters are given in Table 3.23.

The electron density distributions of samples A, BM, BMA are mapped using the electron density values obtained through refinements. Figure 3.29a–f shows the three-dimensional electron densities of samples A, BM, BMA in the unit cell in two different orientations.

Figure 3.30a–c shows the two-dimensional electron densities on (100) plane in the unit cell for the three samples respectively.

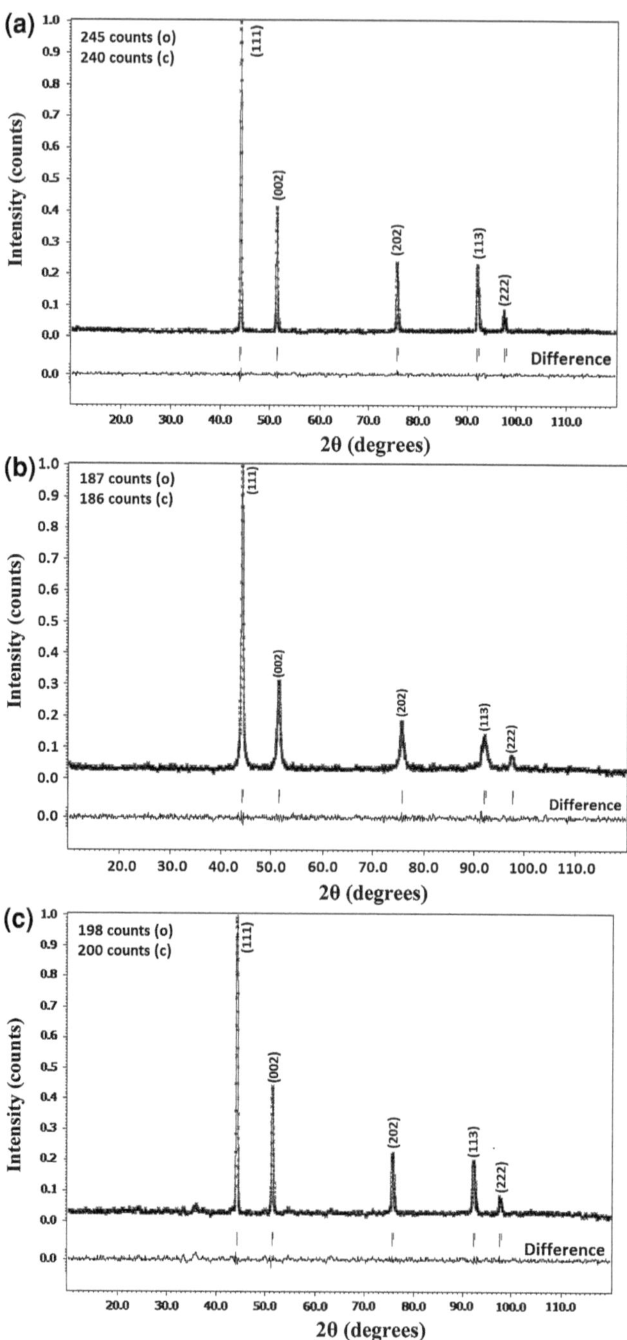

Fig. 3.28 **a** The fitted XRD profiles and the positions of Bragg peaks for sample A. **b** The fitted XRD profiles and the positions of Bragg peaks for sample BM. **c** The fitted XRD profiles and the positions of Bragg peaks for sample BMA

Table 3.21 Observed and calculated structure factors for **a** Sample A, **b** Sample BM, **c** Sample BMA

hkl	Fo	Fc	Fo − Fc	σ (Fo)
a Sample A				
1 1 1	65.62	65.67	−0.05	0.94
0 0 2	59.48	59.67	−0.19	0.93
2 0 2	43.77	42.81	0.96	0.78
1 1 3	34.36	34.60	−0.24	0.79
22 2	31.55	32.40	−0.85	1.31
b Sample BM				
111	65.10	65.15	−0.06	0.93
0 02	59.01	59.09	−0.08	1.11
202	41.87	42.07	−0.20	1.03
113	33.74	33.80	−0.06	0.86
22 2	30.75	31.59	−0.84	1.59
c Sample BMA				
111	66.85	67.11	−0.26	0.96
00 2	61.16	61.43	−0.27	1.21
202	45.17	45.37	−0.20	1.11
113	37.33	37.49	−0.16	0.92
222	34.99	35.36	−0.37	1.68

Table 3.22 The refined structural parameters using JANA 2006

Sample	Cell parameter (Å)	B (Å^2)	R_{obs} (%)	wR_{obs} (%)	Rp (%)	wRp (%)
A	3.5462 (0000)	0.991(017)	0.98	0.12	8.35	11.48
BM	3.5534 (0012)	1.088(256)	0.54	0.54	8.17	11.00
BMA	3.5432 (0013)	0.681(000)	0.51	0.46	10.16	13.72

B Debye–Waller factor

Table 3.23 Refined parameters from MEM technique

Sample	Parameters				
	Number of cycles	Prior density $\tau(r_i)$ (e/Å^3)	Lagrange parameter (λ)	R_{MEM} (%)	wR_{MEM} (%)
A	52	2.44	0.021	1.26	1.55
BM	32	2.43	0.026	0.98	1.09
BMA	8	2.45	0.135	1.86	1.90

Figure 3.31a–c shows the two-dimensional electron densities on (110) plane in the unit cell. The one-dimensional electron density profiles along [100], [110] and [111] directions are represented in Fig. 3.32a–c. Table 3.24 gives the numerical values of the electron densities along the different directions.

The MEM electron density distribution map of sample BM given in Fig 3.30b reveals the core atom to be spherical and shows the enhancement of the electron

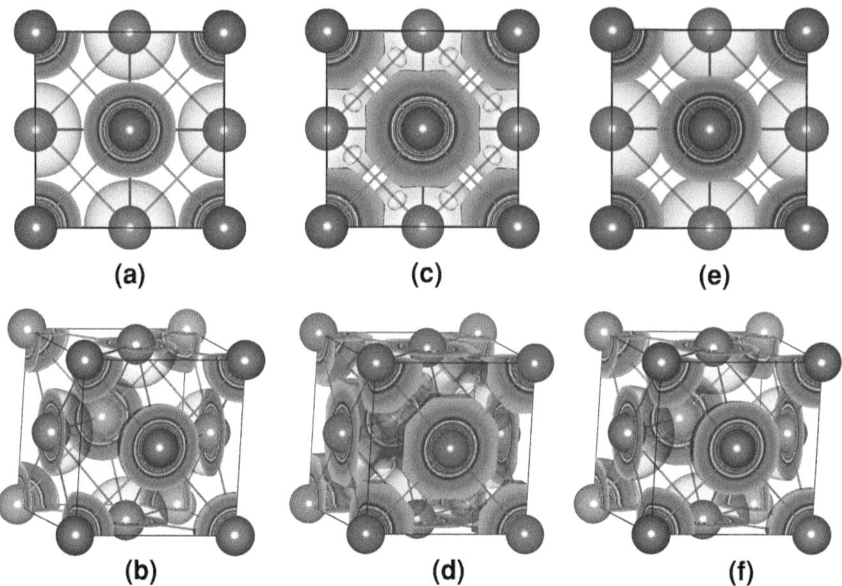

Fig. 3.29 **a**, **b** Three-dimensional MEM electron density isosurface of sample A in two different orientations. **c**, **d** Three-dimensional MEM electron density isosurface of sample BM in two different orientations. **e**, **f** Three-dimensional MEM electron density isosurface of sample BMA in two different orientations

density in sample BM. This enhancement reflects the high thermal vibration in this sample. The two-dimensional MEM electron density distributions on (100) plane shown in Fig. 3.30a–c exhibit the large thermal vibration in sample BM. This is observed from the smearing of charges at the corner atoms and valence electrons in the face-centred atoms. The thermal vibration is comparatively less in sample A. Expansion of the charge region is observed in sample BMA. Expansion of charge clouds of the face-centred atom and their extension towards the corner atoms is observed. This reveals the restriction in the thermal vibration resulting in the low Debye–Waller factor in this case. The increase in the atomic sizes can also be visualized. Ball milling appears to have increased the average cell constant, which is then reduced by subsequent annealing of the ball-milled sample.

The one-dimensional electron density profiles shown in Fig. 3.31a along [100] direction illustrates that the electron density is higher for sample BM initially but at the mid-bond region the density for sample A is increased. This can be attributed to the removal of dislocations and vacancies due to annealing. Along the [110] direction (Fig. 3.32b), which is the bonding direction, the electron density of sample BM is observed to be higher. Ball milling has effectively increased the bond strength in $Ni_{80}Cr_{20}$ along the bonding direction than in the other directions. The flattening of the particles as seen from SEM picture (Fig. 3.26b) is in support of this conclusion. Again, the electron density along [111] direction (Fig. 3.32c),

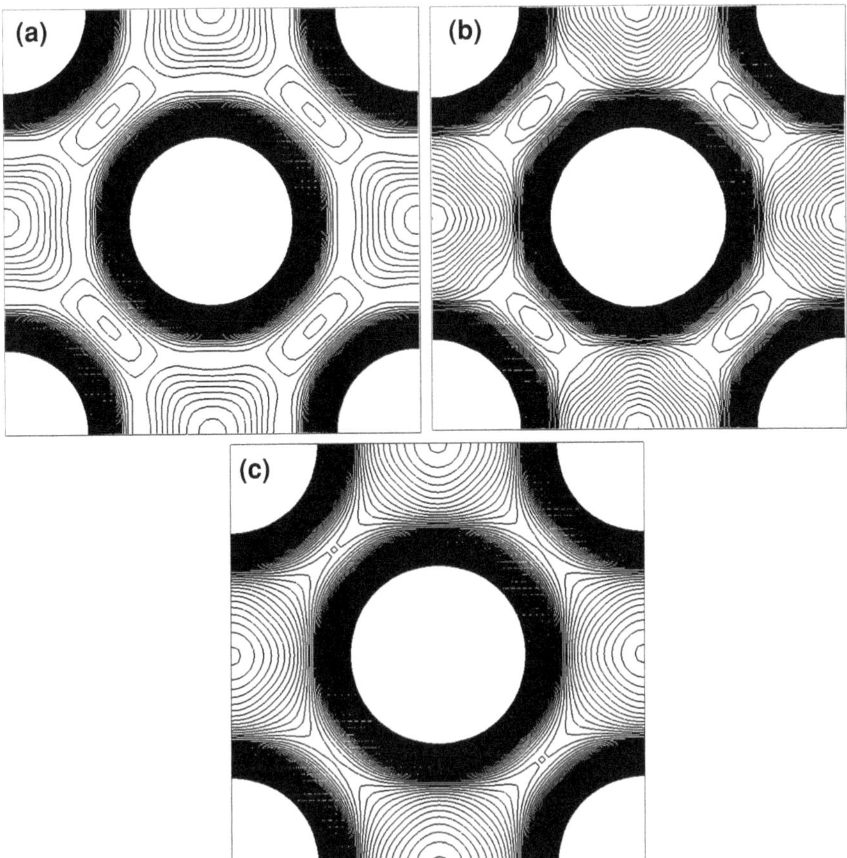

Fig. 3.30 **a** Two-dimensional electron density distributions on the (100) plane for sample A. *Contour lines* are drawn between 0.0 and 1.5 e/Å^3 with 0.1 e/Å^3 interval. **b** Two-dimensional electron density distribution on the (100) plane for sample BM. *Contour lines* are drawn between 0.0 and 1.5 e/Å^3 with 0.1 e/Å^3 interval. **c** Two-dimensional electron density distribution on the (100) plane for sample BM. *Contour lines* are drawn between 0.0 and 1.5 e/Å^3 with 0.1 e/Å^3 interval

show that the density is higher for sample A. In general, sample BMA has lower electron density. Ball milling and subsequent annealing has reduced the charge density in the alloy Ni$_{80}$Cr$_{20}$. The strain produced in the sample BM (as evidenced from the flat region in electron density in Fig. 3.32b) is relieved along the other directions, particularly the [111] direction as seen from Fig. 3.32c. This figure shows undulations in the ball-milled sample which is much higher than the other two samples.

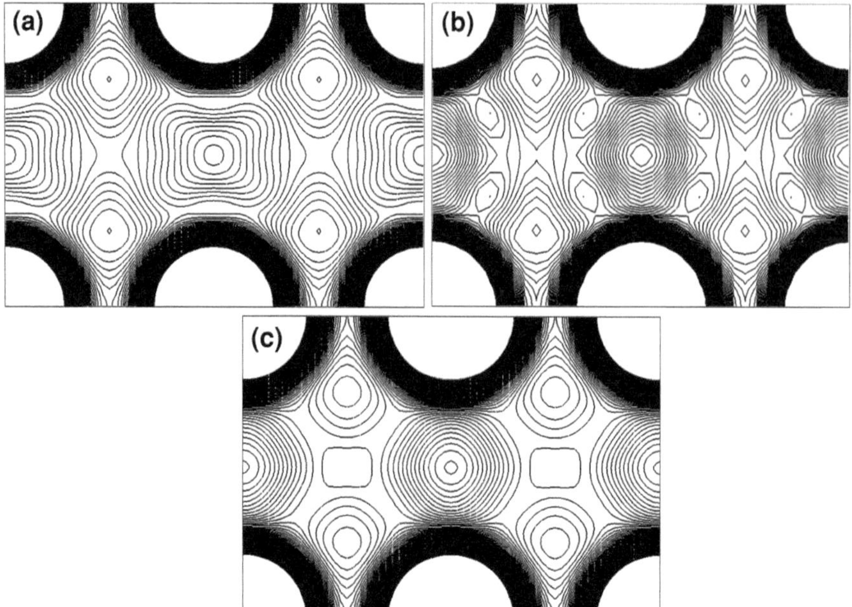

Fig. 3.31 **a** Two-dimensional electron density distribution on (110) plane for sample A. *Contour lines* are drawn between 0.05 and 1.0 e/Å^3 with 0.05 e/Å^3 interval. **b** Two-dimensional electron density distribution on (110) plane for sample BM. *Contour lines* are drawn between 0.05 and 1.0 e/Å^3 with 0.05 e/Å^3 interval. **c** Two-dimensional electron density distribution on (110) plane for sample BMA. *Contour lines* are drawn between 0.05 and 1.0 e/Å^3 with 0.05 e/Å^3 interval

3.5.4.3 Pair Distribution Function

In this study, the observed PDF was obtained from raw X-ray powder data using software programme PDFgetX (Jeong et al. 2001). A comparison between observed and calculated PDFs has been carried out using software package PDFFIT (Farrow et al. 2007). The results of the PDF studies are presented in Table 3.25.

The refined pair distribution function given in Fig. 3.33a–c for samples A, BM, BMA respectively, show good matching of observed and calculated PDFs. The PDF refinement can be considered as equivalent to matching the observed X-ray powder data with too little structure parameters. Moreover, high Q data are required for these types of analyses on short-range order and local structure of materials. Powder XRD data have been used in this work only for comparison purposes.

The bond lengths deduced from the PDF analysis have been compared with those calculated using software programme GRETEP is a Windows interactive programme that displays the list of calculated bonds, bond length between two atoms and the distances and angles around an atom) (Laugier and Bochu). On comparing the observed bond lengths obtained from the PDFs of the three samples, it is found that maximum deviation in the observed bond lengths is noted in sample BM.

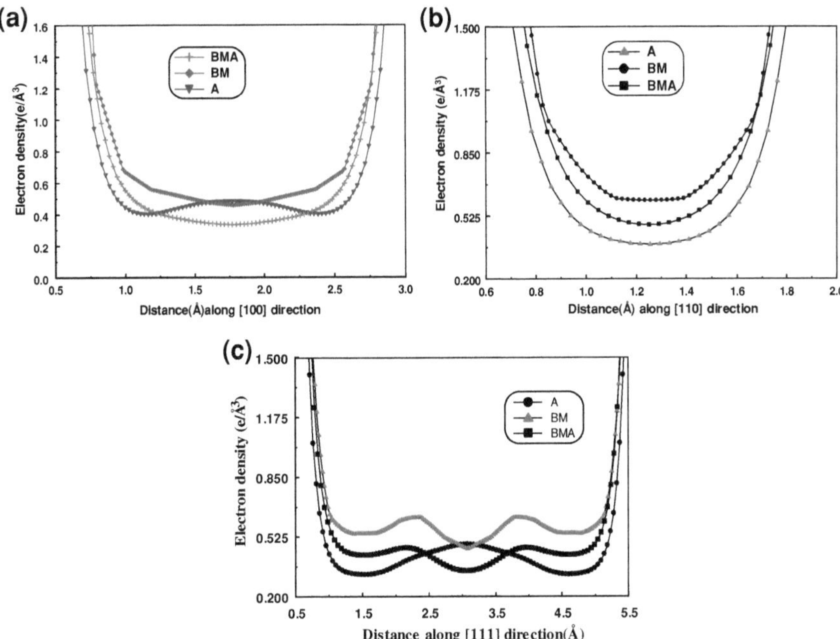

Fig. 3.32 a One-dimensional MEM electron density profiles of samples A, BM,BMA along [100] direction. **b** One-dimensional MEM electron density profiles of samples A, BM, BMA along [110] direction. **c** One-dimensional MEM electron density profiles of samples A, BM, BMA along [111] direction

Table 3.24 Electron density profiles along the three directions of the unit cell

Sample	[100]		[110] (Bonding direction)		[111]	
	Position (Å)	ED (e/Å^3)	Position (Å)	ED (e/Å^3)	Position (Å)	ED (e/Å^3)
A	1.7731	0.484	1.2537	0.378	3.0711	0.484
BM	1.7767	0.460	1.2561	0.605	3.0774	0.460
BMA	1.7691	0.337	1.2509	0.479	3.0641	0.337

ED Electron density

Table 3.25 Neighbour distances from PDF Analysis

Neighbour distance (from PDF analysis) (Å)

Sample	Observed	Calculated[a]
A	2.480	2.508
BM	2.440	2.513
BMA	2.460	2.502

[a] GRETEP software

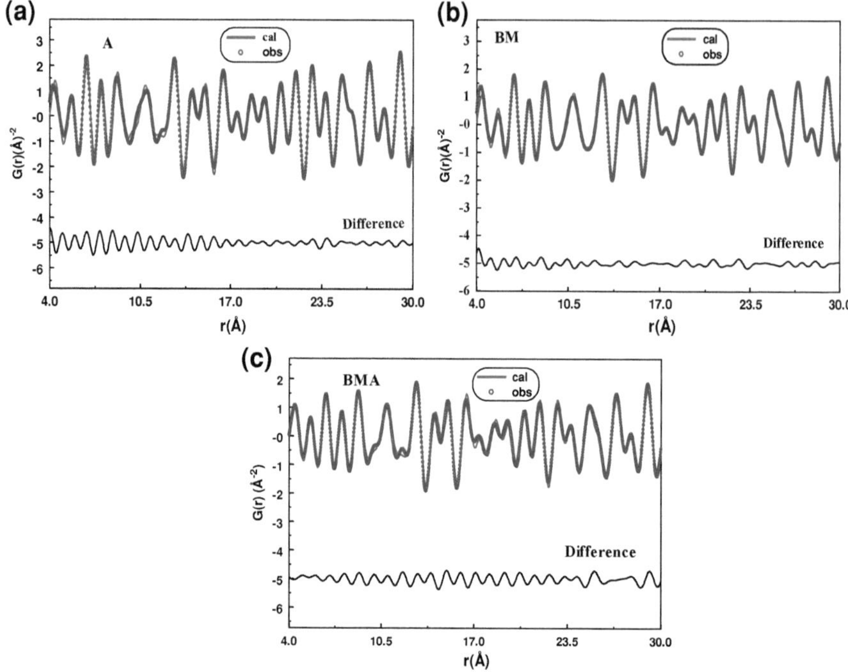

Fig. 3.33 **a** The observed and calculated pair distribution function of sample A. **b** The observed and calculated pair distribution function of sample BM. **c** The observed and calculated pair distribution function of sample BMA

3.5.5 Conclusion

The effects of ball milling and annealing on the local structure have been studied in this work. The ball-milled sample shows enhanced electron density along bonding direction, which decreases on annealing. The cell parameters and the bond lengths appear to be larger when the sample is ball-milled, but after annealing these parameter values are reduced due to relieving of strain in the sample.

3.6 Silver Doped in NaCl ($Na_{1-x}Ag_xCl$)

3.6.1 Introduction

The distribution of electrons in the crystalline system is an interesting and important subject of research in the field of physics, chemistry, material science and Engineering. Particularly in doped systems this study attains great importance to understand the changes in the microscopic and macroscopic properties due to doping. Many studies have been conducted on alkali halides, for example, Linkoaho (1969), Banerjee and Mitra (1969), Shahi and Wagner (1981), Valvoda (1970), Burley (1967). The effect of doping on the physical properties of alkali halides is a subject of sustained interest. While the changes in electrical and optical properties accompanying doping of alkali halides crystals have received considerable attention, lattice strains and lattice deformations arising out of the introduction of foreign atoms have not been studied extensively. Metals containing salts are among the inorganic substances that act as antimicrobial agents. Also, the metal salts are used in gradient index of optical fibre communication, development of passive fault current limiter in parallel biassing, mass spectroscopic analysis of metals etc. The present analysis is carried out on the accurate electron density distribution using MEM and multipole analysis. This research work is on single crystal data sets of $Na_{0.97}Ag_{0.03}Cl$ and $Na_{0.90}Ag_{0.10}Cl$ (Mohanlal and Pathinettam Padiyan 1989). When the system is dilute doped, it is possible to study only the average electronic structure. But, when suitable tools like MEM and multipole techniques are available, they can be used effectively in conjunction to elucidate the fine features of bonding as done in this work. In order to elucidate the valence electrons distribution, MEM and to analyse the contraction/expansion of atomic shells, multipole analysis of the electron density was carried out.

3.6.2 Summary of the Work

The alkali halide $Na_{1-x}Ag_xCl$, with two different compositions (x = 0.03 and 0.10), was studied with regard to the Ag impurities in terms of the bonding and electron density distribution. X-ray single crystal data sets have been used for the purpose. The present analysis focused on the electron density distribution and hence the interaction between the atoms is clearly revealed by MEM and multipole analyses. The bonding in these systems has been studied using two-dimensional MEM electron density maps on the (100) and (110) planes and one-dimensional electron density profiles along the [100], [110] and [111] directions. The mid-bond electron densities between atoms in these systems are found to be 0.175 and 0.183 $e/Å^3$, respectively, for $Na_{0.97}Ag_{0.03}Cl$ and $Na_{0.90}Ag_{0.10}Cl$. Multipole analysis of the structure has been performed for these two systems, with respect to the expansion/contraction of the ion involved.

3.6.3 Data Analysis

The two data sets ($Na_{0.97}Ag_{0.03}Cl$ and $Na_{0.90}Ag_{0.10}Cl$) reported by Mohanlal and Pathinettam Padiyan (1989) have been analysed using Rietveld (1969) structural refinement using the software package Jana 2000 (Petříček et al. 2000). The refined multipole structure factors were used for the MEM computations of the

Table 3.26 Parameters from Rietveld analysis using JANA 2000

Parameter	System	
	$Na_{0.97}Ag_{0.03}Cl$	$Na_{0.90}Ag_{0.10}Cl$
B_{Na} ($\mathring{A}^2$)	2.5208(2296)	2.5284(1483)
B_{Cl} ($\mathring{A}^2$)	1.6912(0959)	1.6251(0774)
R (%)	2.20	2.20
wR (%)	2.42	2.31

B Debye–Waller factor

electron densities. Table 3.26 represents the results of the Rietveld refinements for $Na_{0.97}Ag_{0.03}Cl$ and $Na_{0.90}Ag_{0.10}Cl$ respectively.

3.6.4 MEM Refinement

MEM is one of the appropriate methods in which the concept of entropy is introduced to handle the uncertainty in the electron density distribution properly. The principle of electron density distribution is to obtain results consistent with the observed structure factors and to reduce the uncertainties as much as possible. In this work, MEM is adopted for the refinements of electron densities $Na_{0.97}Ag_{0.03}Cl$ and $Na_{0.90}Ag_{0.10}Cl$, in which maximisation of the initially assumed electron density distribution through refinements and hence the building up of the real electron density distribution has been done. For numerical computations, the software package PRIMA (Izumi and Dilanian 2002) was used. For three-, two- and one-dimensional representation of the electron densities, the programme VESTA (Momma and Izumi 2006) software package was used.

MEM refinements were carried out by dividing the unit cell into $64 \times 64 \times 64$ pixels. The initial electron density at each pixel is fixed uniformly as $F_{000}/a_0^3 = 0.6455$ and 0.7045 e/$\mathring{A}^3$ for $Na_{0.97}Ag_{0.03}Cl$ and $Na_{0.90}Ag_{0.10}Cl$ respectively, where F_{000} is the total number of electrons in the unit cell and a_0 is the cell parameter. The Lagrange parameter is suitably chosen so that the convergence criterion $C = 1$ is reached after the minimum number of iterations. The MEM parameters are given in Table 3.27.

The three-dimensional electron density distribution in the form of isosurface in the unit cell for $Na_{0.97}Ag_{0.03}Cl$ and $Na_{0.90}Ag_{0.10}Cl$ is represented in Fig. 3.34a, b for $Na_{0.97}Ag_{0.03}Cl$ and $Na_{0.90}Ag_{0.10}Cl$ respectively.

The two-dimensional electron density distribution on the (100) and (110) planes is given in Figs. 3.35a, b and 3.36a, b for $Na_{0.97}Ag_{0.03}Cl$ and $Na_{0.90}Ag_{0.10}Cl$ respectively.

The one-dimensional electron density profiles along [100], [110] and [111] directions are represented in Fig. 3.37a, b for these two systems.

Figure 3.38a gives the peak electron density profiles of Na(Ag) along the [100] direction for the two compositions, ($Na_{0.97}Ag_{0.03}Cl$ and $Na_{0.90}Ag_{0.10}Cl$) and Fig. 3.38b gives the mid-bond electron density of Na (Ag) along the [100] direction for

Table 3.27 Parameters from MEM analysis

Parameter	System	
	$Na_{0.97}Ag_{0.03}Cl$	$Na_{0.90}Ag_{0.10}Cl$
Number of cycles	63	56
Prior density, $\tau(r)$, (e/Å^3)	0.646	0.705
Lagrange parameter (λ)	0.279	0.053
RMEM (%)	2.394	2.405
wR MEM (%)	3.001	2.736
Resolution (Å/pixel)	0.0881	0.0880

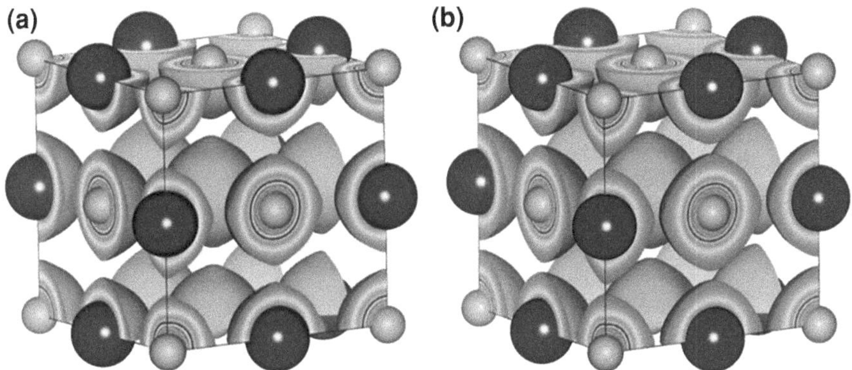

Fig. 3.34 **a** Three-dimensional electron density of $Na_{0.97}Ag_{0.03}Cl$. **b** Three-dimensional electron density of $Na_{0.90}Ag_{0.10}Cl$

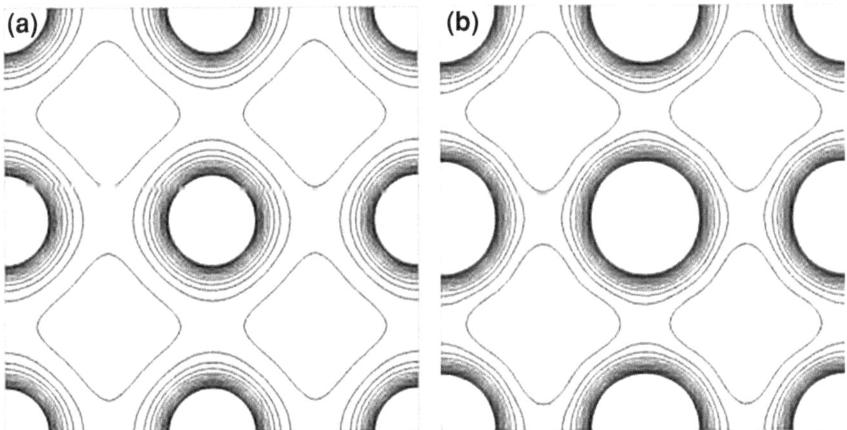

Fig. 3.35 **a** MEM electron density distribution of $Na_{0.97}Ag_{0.03}Cl$ on the (100) plane. **b** MEM electron density distribution of $Na_{0.90}Ag_{0.10}Cl$ on the (100) plane

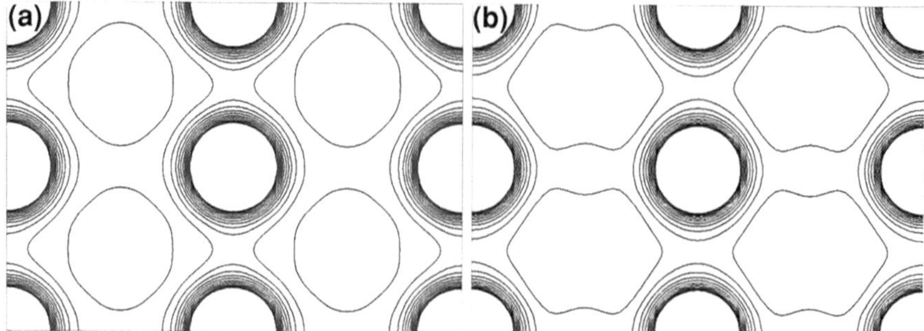

Fig. 3.36 a MEM electron density distribution of $Na_{0.97}Ag_{0.03}Cl$ on the (110) plane. **b** MEM electron density distribution of $Na_{0.90}Ag_{0.10}Cl$ on the (110) plane

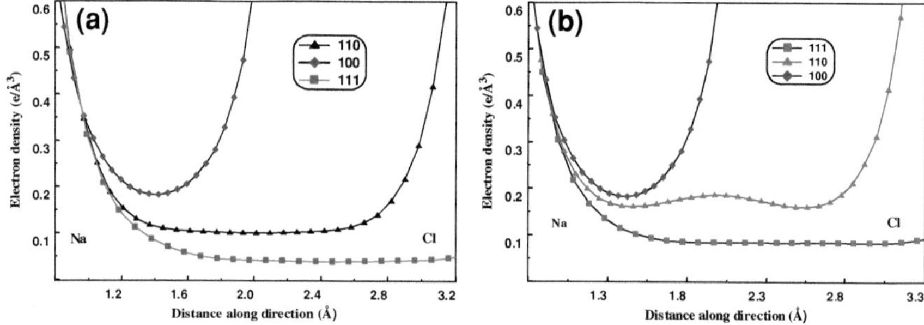

Fig. 3.37 a One-dimensional variation of electron density along [100], [110] and [111] directions of the $Na_{0.97}Ag_{0.03}Cl$ unit cell. **b** One-Dimensional variation of electron density along [100], [110] and [111] directions of the $Na_{0.90}Ag_{0.10}Cl$ unit cell

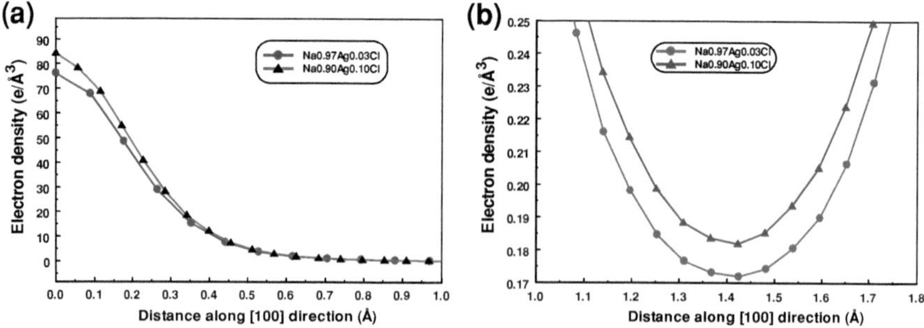

Fig. 3.38 a The peak electron density profiles of Na (Ag) along the [100] direction for the two compositions, ($Na_{0.97}Ag_{0.03}Cl$ and $Na_{0.90}Ag_{0.10}Cl$). **b** The mid-bond electron density of Na (Ag) along the [100] direction for $Na_{0.97}Ag_{0.03}Cl$ and $Na_{0.90}Ag_{0.10}Cl$

Table 3.28 Parameters from MEM refinement

Direction	Na$_{0.97}$Ag$_{0.03}$Cl		Na$_{0.90}$Ag$_{0.10}$Cl	
	Position (Å)	Electron density (e/Å^3)	Position (Å)	Electron density (e/Å^3)
[001]	1.422	0.175	1.423	0.183
[110]	2.010	0.148	2.012	0.187
[111]	2.462	0.057	1.971	0.084

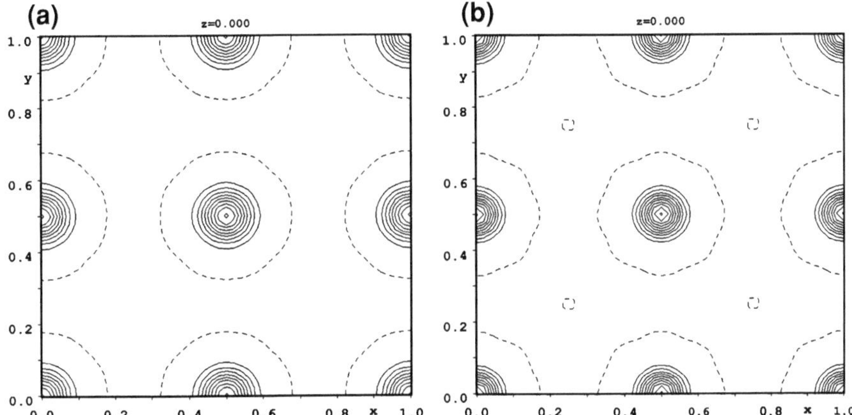

Fig. 3.39 **a** Dynamic multipole deformation (DMD) map of Na$_{0.97}$Ag$_{0.03}$Cl on the (100) plane. **b** Dynamic multipole deformation (DMD) map of Na$_{0.90}$Ag$_{0.10}$Cl on the (100) plane

Na$_{0.97}$Ag$_{0.03}$Cl and Na$_{0.90}$Ag$_{0.10}$Cl. The numerical values of the electron densities along different directions are given in Table 3.28.

3.6.5 Multipole Analysis

Hansen and Coppens (1978) proposed a modified electron density model with the option that allows the refinement of population parameters at various orbital levels where the atomic density is described as a series expansion in real spherical harmonic functions through fourth order Y$_{lm}$.

Multipole electron density computation was carried out according to Hansen and Coppens (1978) in this work. The electron density from multipole analysis is represented in Fig. 3.39a, b as DMD maps on the (100) plane for Na$_{0.97}$Ag$_{0.03}$Cl and Na$_{0.90}$Ag$_{0.10}$Cl respectively.

Figure 3.40a, b represent the SMD maps on (100) plane for Na$_{0.97}$Ag$_{0.03}$Cl and Na$_{0.90}$Ag$_{0.10}$Cl respectively.

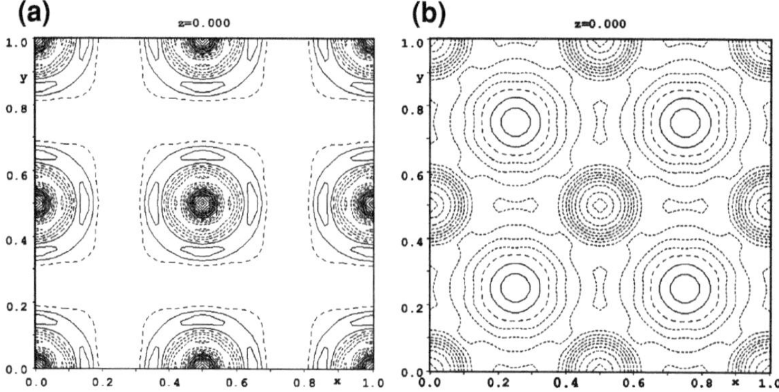

Fig. 3.40 **a** Static multipole deformation map of $Na_{0.97}Ag_{0.03}Cl$ on the (100) plane. **b** Static multipole deformation map of $Na_{0.90}Ag_{0.10}Cl$ on the (100) plane

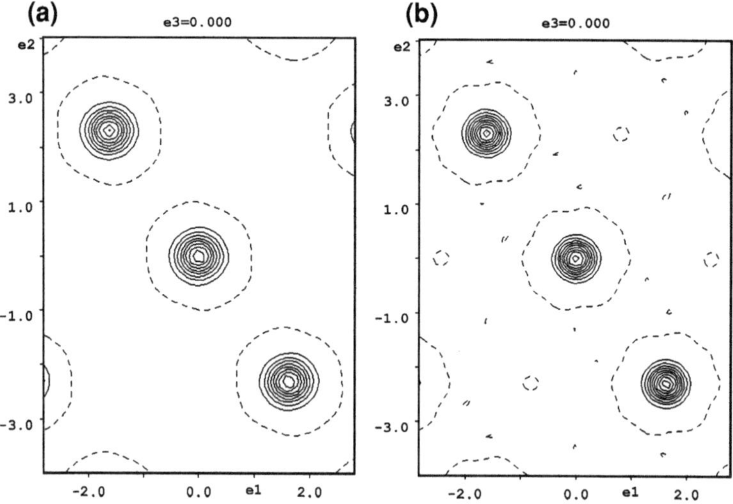

Fig. 3.41 **a** Dynamic multipole deformation map of $Na_{0.97}Ag_{0.03}Cl$ on the (110) plane. **b** Dynamic multipole deformation map of $Na_{0.90}Ag_{0.10}Cl$ on the (110) plane

Table 3.29 Multipole electron densities

System	Residual densities	
	$Na_{0.97}Ag_{0.03}Cl$	$Na_{0.90}Ag_{0.10}Cl$
DMD analysis		
Positive max. (e/Å^3)	1.96	8.7
Negative max. (e/Å^3)	−0.04	−0.18
SMD analysis		
Positive max. (e/Å^3)	0.11	0.11
Negative max.(e/Å^3)	−0.04	−0.11

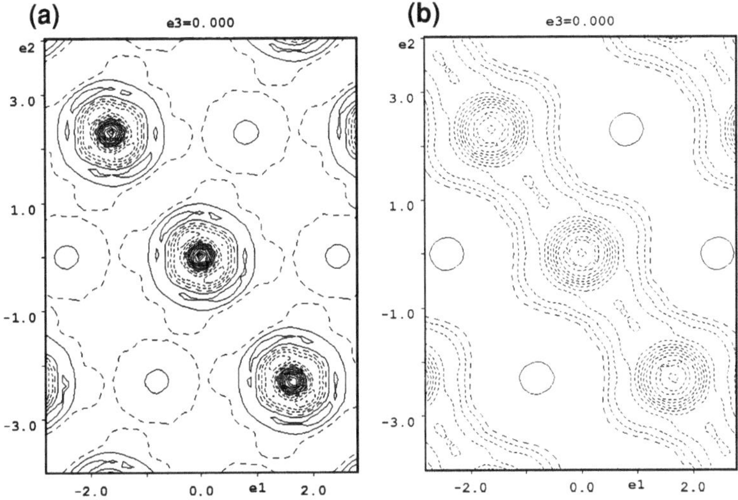

Fig. 3.42 a Static multipole deformation map of $Na_{0.97}Ag_{0.03}Cl$ on the (110) plane. **b** Static multipole deformation map of $Na_{0.90}Ag_{0.10}Cl$ on the (110) plane

Figure 3.41a, b represent DMD maps on (110) plane for $Na_{0.97}Ag_{0.03}Cl$ and $Na_{0.90}Ag_{0.10}Cl$ respectively.

Similarly Fig. 3.42a, b represent the SMD maps on (110) plane for $Na_{0.97}Ag_{0.03}Cl$ and $Na_{0.90}Ag_{0.10}Cl$ respectively. The results of the multipole analysis are given in Table 3.29.

3.6.6 Results and Discussion

The effect of impurity doping has been analysed in this work, in terms of the electron density distribution using the two versatile techniques MEM and multipole analysis. The experimental structure factor data sets of $Na_{0.97}Ag_{0.03}Cl$ and $Na_{0.90}Ag_{0.10}Cl$ were subjected to Rietveld (1969) refinements to obtain the individual thermal vibration parameters of the atoms Na(Ag) and Cl. The refined thermal vibration parameters as given in Table 3.26 shows that the Debye–Waller factor of Na(Ag) increases with the concentration of Ag, whereas the Debye–Waller factor value of Cl decreases. When the NaCl is doped with Ag impurity, the local distortion increases with impurity concentration. Hence, the Debye–Waller factor of Na(Ag) increases with Ag. Since the electrostatic interaction between the Na(Ag) and Cl ions increases with Ag concentration, the Cl atom has less thermal vibration as reflected in Table 3.26. The Rietveld refinements give very low reliability indices as given in Table 3.26. The refined results were used for the MEM computations. The three-Dimensional electron density of $Na_{0.97}Ag_{0.03}Cl$ and $Na_{0.90}Ag_{0.10}Cl$ in terms of isosurfaces as shown in Fig. 3.34a, b can be visually compared for the ionic sizes and the interionic distances. It is seen from these figures that the electronic charge clouds are more diffuse in $Na_{0.90}Ag_{0.10}Cl$ and the separation between charges decreases as the concentration of Ag increases. Since the charge clouds of Cl expand due to more electrostatic attraction in $Na_{0.90}Ag_{0.10}Cl$, the thermal vibration of Cl is restricted, resulting in smaller Debye–Waller factor of Cl.

The increase in the ionic sizes can be visualized in the two-Dimensional electron density contour maps on (100) and (110) plane (Figs. 3.35a, b and 3.36a, b) for $Na_{0.97}Ag_{0.03}Cl$ and $Na_{0.90}Ag_{0.10}Cl$ respectively. Particularly, Fig. 3.36b for $Na_{0.90}Ag_{0.10}Cl$ shows increased ionic sizes compared to those in Fig. 3.36a. Hence, the effect of impurity addition in the host lattice is substantiated.

From the one-dimensional analysis of electron density it is found that there is an increase in the mid-bond density in $Na_{0.90}Ag_{0.10}Cl$ compared to that in $Na_{0.97}Ag_{0.03}Cl$ in all the three directions. The minimum density along [100] direction occurs at 1.4251 and 1.4227 Å respectively, for $Na_{0.97}Ag_{0.03}Cl$ and $Na_{0.90}Ag_{0.10}Cl$ indicating an increase in the ionic radius in $Na_{0.90}Ag_{0.10}Cl$. Moreover, the electron density at these positions is 0.175 and 0.183 $e/Å^3$ for $Na_{0.97}Ag_{0.03}Cl$ and $Na_{0.90}Ag_{0.10}Cl$ respectively. This clearly indicates the addition of Ag ion in the host lattice. A similar trend i.e., increased mid-bond position and density is observed along [110] direction also, in $Na_{0.90}Ag_{0.10}Cl$. Along [111] direction, there is an increase in electron density in $Na_{0.90}Ag_{0.10}Cl$ compared to $Na_{0.97}Ag_{0.03}Cl$, but the position of mid-bond density is decreased in $Na_{0.90}Ag_{0.10}Cl$. But one can observe horizontal lines along [111] direction and only the minimum numerical value decides the electron density, but the bonding characteristics is decided by the electron density distribution along [100] direction which supports the addition of Ag impurities in the host lattice. The [100] profiles provide an estimation of the ionic radii of atoms Na(Ag) and Cl.

The DMD maps on (100) plane (Fig. 3.39a, b) for $Na_{0.97}Ag_{0.03}Cl$ and $Na_{0.90}Ag_{0.10}Cl$ clearly indicate that the residual densities are more concentrated and contracted in $Na_{0.90}Ag_{0.10}Cl$ than $Na_{0.97}Ag_{0.03}Cl$. The SMD maps in Fig. 3.40a, b show positive clusters (contour line) between the atom indicative of more electrostatic interaction in $Na_{0.90}Ag_{0.10}Cl$ compared to $Na_{0.97}Ag_{0.03}Cl$. A similar trend is observed on (110) planes (Fig. 3.41a, b). Figure 3.42b shows (for $Na_{0.90}Ag_{0.10}Cl$) charge clouds of larger radius between the atoms compared to the electron clouds of $Na_{0.90}Ag_{0.10}Cl$. The multipole analysis gives the multipole parameters κ and κ' which are as follows. The κ and κ' values of Na(Ag) in $Na_{0.97}Ag_{0.03}Cl$ and $Na_{0.90}Ag_{0.10}Cl$ was refined keeping the Cl ion fixed. For $Na_{0.90}Ag_{0.10}Cl$ the κ and κ' parameters are 1.805(0.605) and 1.0703(0.5411) respectively. These values are 1.776(0.5040) and 1.0009(0.0544) respectively for $Na_{0.90}Ag_{0.10}Cl$. Both values for $Na_{0.90}Ag_{0.10}Cl$ reveal an expansion of Na(Ag) ion in $Na_{0.90}Ag_{0.10}Cl$. Table 3.29 shows that the residual density is larger in $Na_{0.90}Ag_{0.10}Cl$ as analysed by DMD method.

3.6.7 Conclusion

Both MEM and multipole analyses give a clear picture of the bonding in the doped system NaCl. Hence, these two methods can be used in combination to study the electronic charge distribution in pure and doped systems which will help to understand the physical and conducting phenomena of these materials.

3.7 Aluminium Doped with Dilute Amounts of Iron Impurities (0.215 and 0.304 wt% Fe)

3.7.1 Introduction

The analysis of bonding and electronic structure of metals is useful because of the variety of applications of metals, particularly when they are doped with other elements. There has been tremendous impetus in the study of electronic structure of solids, both theoretically and experimentally due to necessity. When the system is dilute doped, it is possible to study only the average electronic structure. But, when suitable tools like MEM are available, they can be effectively used to elucidate the fine features of bonding as attempted in this work. Fourier synthesis of electron densities can be of use in picturing bonding between two atoms, but, it suffers from the major disadvantage of series termination error and negative electron densities that prevent the clear understanding of the fine nature of bonding between atoms, the factor intended to be analysed. The advent of MEM solves many of these problems. MEM electron densities are always positive and even with limited data, one can determine reliable electron densities resembling true densities. In this work, such an analysis has also been carried out using the formalism of Collins (1982) using the data sets reported by Mohanlal and Pathinettam Padiyan (1987).

In our series of research studies on bonding in materials, we have analysed LiF, NaF, GaAs, CdTe and some oxides (Israel et al. 2002; Israel et al. 2003; Saravanan et al. 2002a, b; Kajitani et al. 2001) using MEM and reported X-ray data and obtained precise information about bonding in the above materials. The nature of bonding in these materials as analysed are found to be ionic, mixed covalent and ionic and 'oxide' bonding.

3.7.2 Summary of the Work

The electronic structure of pure and doped aluminium with dilute amounts of iron impurities (0.215 and 0.304 wt% Fe) has been analysed using reported X-ray data sets and MEM. Qualitative as well as quantitative assessment of the electron density distribution in these samples has been made. The mid-bond character-ization leads to a conclusion about the nature of doping of impurities. An expan-sion of size of the host aluminium atom is observed with Fe impurities.

In this study on aluminium with iron impurities, we have attempted to study metal bonding and bring out the details of doping using MEM. The X-ray structure factor data sets have been taken from Mohanlal and Pathinettam Padiyan (1987) who reported precise structure factors collected from a home-built manual X-ray diffractometer. Data sets for pure aluminium, Al + 0.189, Al + 0.215 and Al + 0.304 wt% Fe are available from this report Mohanlal and Pathinettam Padiyan (1987). Other relevant details can be found from this paper. Among these data sets, the one corresponding to Al + 0.189 wt% Fe has been excluded from the present analysis after examining the preliminary results from MEM and multipole methods (the low doping level does not allow any meaningful conclusions of the obtained results). The remaining three data sets, i.e., those of pure aluminium, Al + 0.215 and Al + 0.304 wt% Fe were used in this study. It was observed that the B_{eff} and the expected number of vacant sites in Al + 0.215 wt% Fe are much better than those in Al + 0.189 wt% Fe, although slight variations of these quantities is observed, compared to pure aluminium and Al + 0.304 wt% Fe [it has been reported (Mohanlal and Pathinettam Padiyan (1987) that the number of vacant sites created in the Al lattice increases with the number of Fe impurities and in the case of Al + 0.189 wt% Fe, less vacancies are created than expected]. Hence, a data set for Al + 0.215 wt% Fe has been included in the present analysis in order to have a composition of iron between pure aluminium and Al + 0.304 wt% Fe and to compare the results.

3.7.3 Data Analysis

All the reported data sets have been analysed using JANA 2000 (Petříček et al. 2000) software package. The refined structure factors were used for MEM refinements of the electron densities. The characteristics of the structural refine-ments are given in Table 3.30.

The reliability indices (R-factors) and the matching of thermal parameters (B) with the corresponding reported values for all three crystal structures indicate that the quality of the experimental data is suitable enough for the electron density analysis. The same number of data for all three compositions samples has been utilised for the present analysis in order to make a comparison of the results.

MEM calculations were done by dividing the cell into 128^3 pixels. For numerical MEM computations, the software package PRIMA (Izumi and Dilanian 2002) was used. For two- and three-dimensional representations of the electron densities, the programme VESTA (Momma and Izumi 2006) package was used. In all three cases, the convergence condition $C = 1.0$, was obtained after several cycles of MEM refinement. The main characteristics of the refinements are listed in Table 3.31.

Table 3.30 Structure refinement parameters from JANA 2000

Sample	Cell constant (Å)	Number of reflections	B (Å^2)	R (%)	wR (%)
Al	4.0440	18	0.825(069)	2.22	2.82
Al + 0.215% Fe	4.0501	18	1.038(061)	2.61	2.86
Al + 0.304% Fe	4.0452	18	1.442(053)	2.61	2.82

B Debye–Waller factor

Table 3.31 Parameters from MEM refinements

Sample	Number of reflections	Cycles	R(%)	wR(%)	Lagrange parameter (λ)	Resolution of the map (Å/pixel)
Al	18	713	0.2276	0.2867	0.010	0.0316
Al + 0.215% Fe	18	832	0.2939	0.2969	0.012	0.0316
Al + 0.304% Fe	18	5703	0.2673	0.3189	0.012	0.0316

Table 3.32 Maxima of the MEM electron density revealed in different one-dimensional sections of a unit cell for different compositions

Composition	Direction					
	[100]		[110]		[111]	
	Position (Å)	ED (e/Å^3)	Position (Å))	ED (e/Å^3)	Position (Å)	ED (e/Å^3)
Al	0.000[a]	128.75	0.000[a]	128.75	0.000[a]	128.75
	1.485	0.14	1.429	0.22	1.477	0.15
	2.022	0.19	1.698	0.20	1.751	0.15
Al + 0.215% Fe	0.000[a]	109.21	0.000[a]	109.21	0.000[a]	109.21
	1.455	0.14	1.298	0.19	1.534	0.15
	2.025	0.19	1.432	0.19	1.753	0.15
					2.631	0.12
					3.508	0.19
					4.384	0.12
					5.261	0.15
Al + 0.304% Fe	0.000[a]	85.79	0.000[a]	85.79	0.000[a]	85.79
	1.517	0.12	1.430	0.15	1.752	0.11
	2.022	0.15			2.080	0.11
					2.682	0.09
					3.503	0.15
					4.324	0.09
					5.254	0.11

[a] At origin

ED Electron density

The reliability indices are found to be very low. To analyse the MEM results, one-dimensional profiles along different directions of the unit cell and two-dimensional electron density contour maps have been constructed. The results of the one-dimensional analysis have given in Table 3.32.

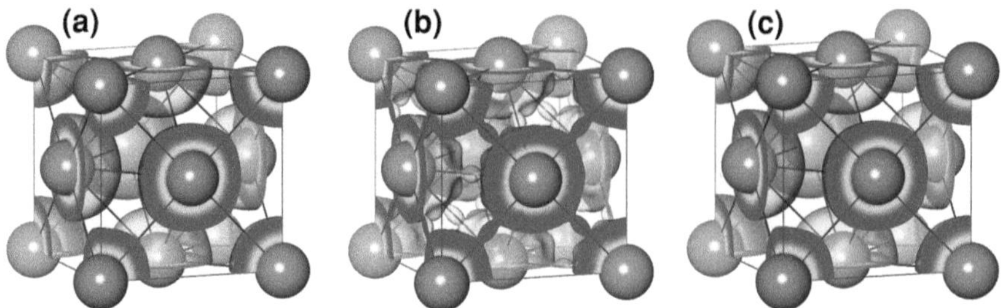

Fig. 3.43 a 3-D representation of the electron density of pure Al super imposed with the structural model. Isosurface level is 0.2. **b** 3-D representation of the electron density of Al + 0.215% Fe super imposed with the structural model. Isosurface level is 0.2. **c** 3-D representation of the electron density of Al + 0.304% Fe imposed with the structural super imposed with the structural model. Isosurface level is 0.2

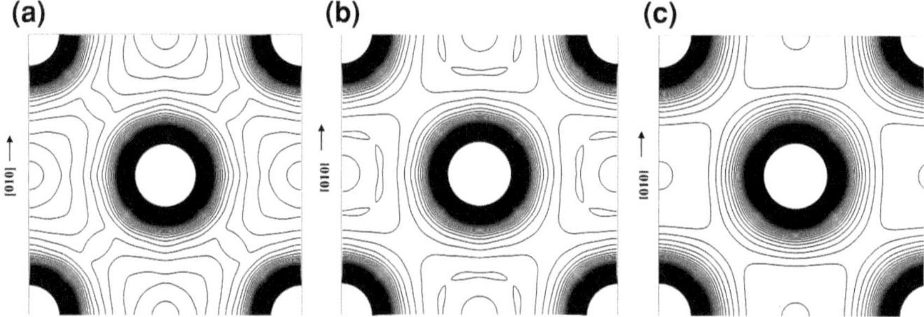

Fig. 3.44 a MEM electron density distribution of pure Al on the (100) plane. Contour range is from 0.0 to 5.0 e/$Å^3$. Contour interval is 0.03 e/$Å^3$. **b** MEM electron density distribution of Al + 0.215% on the (100) plane. Contour range is from 0.0 to 5.0 e/$Å^3$. Contour interval is 0.03 e/$Å^3$. **c** MEM electron density distribution of pure Al + 0.304% on the (100) plane. Contour range is from 0.0 to 5.0 e/$Å^3$. Contour interval is 0.03 e/$Å^3$

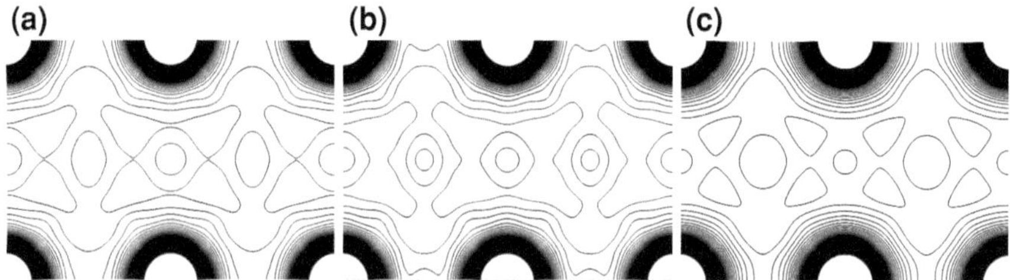

Fig. 3.45 a MEM electron density distribution of pure Al on the (110) plane. Contour range is from 0.0 to 5.0 e/$Å^3$. Contour interval is 0.03 e/$Å^3$. **b** MEM electron density distribution of pure Al + 0.215% Fe on the (110) plane. Contour range is from 0.0 to 5.0 e/$Å^3$. Contour interval is 0.03 e/$Å^3$. **c** MEM electron density distribution of pure Al + 0.304% Fe on the (110) plane. Contour range is from 0.0 to 5.0 e/$Å^3$. Contour interval is 0.03 e/$Å^3$

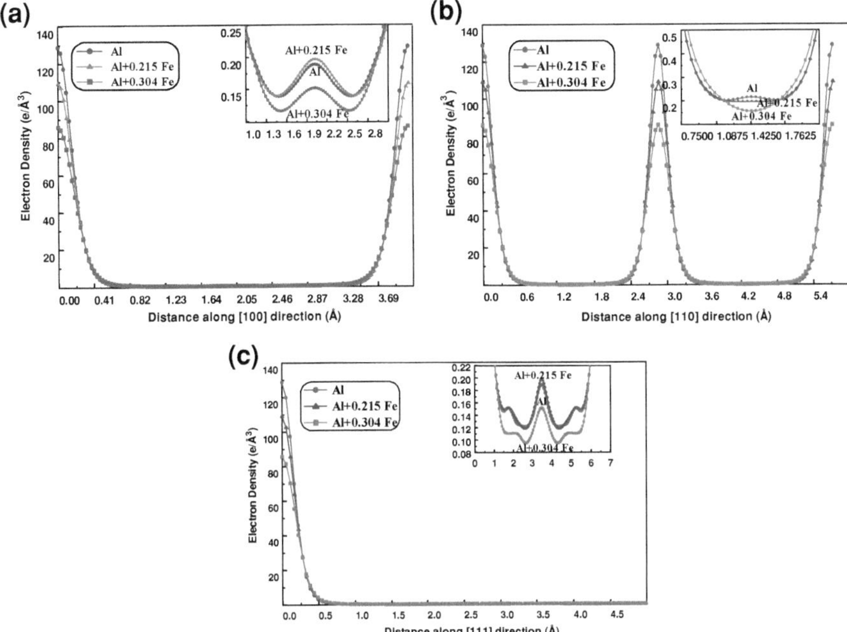

Fig. 3.46 **a** One-dimensional MEM low electron density profiles of pure Al, Al + 0.215% Fe and Al + 0.304% Fe along the [100] direction. The mid-bond variation of the electron density is shown as insert in which the y-axis represents the electron density in e/Å^3 and the x-axis, the distance along the [100] direction in Å. **b** One-dimensional MEM low electron density profiles of pure Al, Al + 0.215% Fe and Al + 0.304% Fe along the [110] direction. The mid-bond variation of the electron density is shown as insert, in which the y-axis represents the electron density in e/Å^3 and the x-axis, the distance along the [110] direction in Å. **c** One-dimensional MEM low electron density profiles of pure aluminium, Al + 0.215% Fe and Al + 0.304% Fe along the [111] direction. The mid-bond variation of the electron density is shown as insert, in which the y-axis represents the electron density in e/Å^3 and the x-axis, the distance along the [111] direction in Å

The three-dimensional electron density superimposed with the structural model of Al is shown for three compositions in Fig. 3.43a, b, c respectively. The 2-D electron density contour maps drawn on (100) planes are given in Fig. 3.44a–c. Figure 3.45a–c represent the 2-D maps on the (110) plane. The one-dimensional variation of the electron density is shown in Fig. 3.46a–c. Figure 3.46a represents the variation along the [100] direction for all three compositions, Fig. 3.46b along the [110] direction and Fig. 3.46c along the [111] direction. The analysis on the MEM results is given in the following section.

3.7.4 Results and Discussion

As expected from the substitution Al by Fe, the peak density decreases with the increase of Fe impurity (Fig. 3.46). As more and more Fe atoms occupy the host aluminium lattice substitutionally, the size of the charge cloud increases due to the larger atomic radius of iron (1.72 Å—for aluminium, the atomic radius is 1.43 Å). The electron density between atoms along [110] direction also decrease with impurities (Table 3.32 and Fig. 3.45b) although it is almost a constant along the other two directions, [100] and [111] (Table 3.32). The bonding direction shows (which is [110] direction, because the first nearest neighbours are along this direction), that the mid-bond electron density also decreases with the addition of Fe impurities as shown in the inset of Fig. 3.46b. The mid-bond density along the other two directions, [100] and [111] (insets of Fig. 3.46a, c), do not reveal any trend with the Fe impurities, because the values of the densities involved are very low along these non-bonding directions and the humps seen in the insets of Fig. 3.46a, c are due to the interference of the electron density of the FCC atoms in the unit cell (NNM—Non Nuclear Maxima, along the non-bonding directions). The MEM electron density distribution of pure Al, Al + 0.215% Fe and Al + 0.304% Fe on the (100) plane (Fig. 3.44a–c) shows clear densities with very low noise levels. The large thermal vibration and smearing of charges is seen. Hence the increased size of the electronic clouds can be attributed to the static and dynamic thermal vibration parameters and also to the increased host lattice substitution of the Fe atoms.

3.7.5 Conclusion

The electron density distribution in the unit cell of Al with different iron impurity has been studied precisely using the currently available most versatile technique, the MEM analysis of the electron densities. Minute changes in the density details and the incorporation of impurities in the host lattice have been revealed. The electron density and the mid-bond electron density decrease with more Fe impurity doping as evidenced from the 1D electron densities. Hence, the bonding strength in Al decreases with Fe addition. In addition to the electron density details showing up themselves as actual distributions of atomic electrons, the thermal parameters also show a similar trend with respect to impurity addition. Moreover, an expansion of the host lattice with the addition of more Fe impurities is observed pictorially as revealed by the two-dimensional electron densities in Fig. 3.44. Since, the crystallographic system presently studied is highly symmetric, there are no appreciable changes in the mid-bond positions despite the increase in the atomic sizes with more addition of Fe impurities.

References

Ak NF, Tekmen C, Ozedmir I, Soykan HS, Celik E (2003) Surf Coat Technol 174:1070–1073
Al-Quran FMF, Al-Itawi HI (2010) Eur J Sci Res 39:251–256
Banerjee NK, Mitra GB (1969) Indian J Pure Appl Phys 71:13
Bashir J, Butt NM, Khan QH (1988) Acta Cryst A44:638–639
Bowran DT, Finney JL (2003) J Chem Phys 118:8357–8372
Burley G (1967) Acta Cryst 1(23):1–5
Collins DM (1982) Nature 298:49–51
Coppens P, Volkov A (2004) Acta Cryst A60:357–364
Debay P, Menki H (1930) Physika Zeit 31:797–798
Egami T (1990) JIM 31(3):163–176
Enríquez JP, Mathew X (2003) J Cryst Growth 259:215–222
Eskandarany MS (2001) Mechanical alloying for fabrication of advanced engineering materials (Chapter 1). Noyes publications, New York
Farrow CL, Juhas P, Liu JW, Bryndin D, Bozin ES, Bloch J, Proffen T, Billinge SJL (2007) PDFfit2 and PDF gui: computer programs for studying nanostructures in crystals. J Phy Cond Matt Phy 19:335219
Field DW, Medlin EH (1974) An X-ray sturdy of thermal vibrations in sodium. Acta Cryst A30:234–238
Frils J, Jiang B, Spence J, Marthinsen K, Holmestad R (2004) Acta Cryst A60:402–408
Gull SF, Daniel GJ (1978) Nature 272:686–690
Hansen NK, Coppens P (1978) Acta Crys A34:909–921
Ikeda T, Takata M, Sakata M, Waliszewski J, Dobrzynski L, Porowski S, Jun J (1998) J Phys Soc Jpn 67:4104–4109
Israel S, Saravanan R, Srinivasan N, Rajaram RK (2002) Phys Chem Solids 64:43–49
Israel S, Saravanan R, Srinivasan N, Rajaram RK (2003) J Phys Chem Solids 64:879–886
Israel S, Saravanan R, Rajaram RK (2004) Physica B 349:390–400
Izumi F, Dilanian RA (2002) Recent research developments in physics, vol 3, Part II. Transworld Research Network, Trivandrum , pp 699–726
Jeng SP, Holloway H, Douglas A, Hoflund GB (1990) Surf Sci 235:175–185
Jeong IK, Thompson J, Proffen T, Perez A, Billinge SJL (2001) PDFGetX: a program for obtaining the atomic pair distribution function from X-ray powder diffraction data. J Appl Cryst 34:536
Kajitani T, Saravanan R, Ono Y, Ohno K, Isshiki M (2001) J Cry Growth 229:130–136
Kamal S, Jayaganthan R, Prakash S (2009) J Alloy Compd 472:378–389
Kazi IH, Wild PM, Moore TN, Sayer M (2003) Thin Solid Films 433:337–343
Kazi IH, Wild PM, Moore TN, Sayer M (2006) Thin Solid Films 515:2602–2606
Kim N, Kwon NH, Park Y, Choi J (2005) Electron Lett 41:982–984
Larson AC, Von Dreele RB (2004) General structure analysis system, Lansce, MS-H805, Los Alamos national laboratory, Los Alamos, LAUR 86748
Linkoaho MV (1969) Acta Crystallographica A 25 III:450–455
Linkoaho MV (1972) Physica Scripta 5:271–272
Marabello D, Bianchi R, Gervasio G, Cargnoni F (2004) Acta Cryst A60:494–501
Mohanlal SK (1979) J Phys C Solid State Phys 12:651–653
Mohanlal SK, Pathinettam Padiyan D (1987) Cryst Latt Def Amorph Mat 14:651–653
Mohanlal SK, Pathinettam Padiyan D (1989) Radiat Eff Defects Solids 110:385–391
Momma K, Izumi F (2006) Commission on crystallography computing. IUCr Newslett 7:106–119
Murray GT (1997) Handbook of materials selection for engineering applications. Dekker, Inc. New York
Na SH, Kim SH, Lee YW, Sohn DS (2002) J Korean Nucl Soc 34:60–67
Peng LM, Ren G, Dudarey SL, Whelan M (1996) J Acta Cryst A52:456–470

Peng Y, Cullis T, Möbus G, Xu X, Inkson B (2009) IOP Electron J Nanotechnology 20: 395708–395714

Peterson PF, Bozin ES, Proffen TH, Billinge SJL (2003) J Appl Cryst 36:53–64

Petkov V, Billinge SJL, Larsen PD, Mahanthi T, Vogt KK, Rangen MG, Kanatzidis (2002) Phys Rev B 65:092105–092108

Petříček V, Dušek M, Palatinus L (2000) JANA 2000: the crystallographic computing system, institute of physics. Academy of Sciences, Praha

Petříček V, Dušek M, Palatinus L (2006) JANA 2000: the crystallographic computing system, institute of physics. Academy of Sciences, Praha

Pillet S, Souhassou M, Lecomte C (2004) Acta Cryst A60:455–464

Poulsen RD, Bentien A, Graber T, Iversen BB (2004) Acta Cryst A60:382–389

Proffen T, Billinge SJL (1999) PDFFIT: a Program for full profile structural refinement of the atomic pair distribution function. J Appl Cryst 32:572–575

Proffen T, Neder R (1997) DISCUS: a Program for diffuse scattering and defect structure simulations. J Appl Cryst 30:171–175

Radomysel'skii ID, Kholodnyi IP (1975) Powder Metall Met Ceram 14(6):478–486

Ram A (1985) CRC handbook of chemistry and physics, 64th edn. CRC press, Florida

Rietveld HM (1969) J Appl Crystallogr 2:65–71

Robert MC, Saravanan R, Saravanakumar K, Prema Rani M (2009) Z Naturforsch 64a:1–9

Ruben AD, Fujio I (2004) Super-fast program PRIMA for the maximum-entropy method, advanced materials laboratory, National Institute for Materials Science, Ibaraki, Japan, p 3050044

Saka T, Kato N (1986) Acta Cryst A 42:469–478

Sakata M, Sato M (1990) Acta Cryst A 46:263–270

Saravanan R, Prema Rani M (2007) J All Compd 431:121–129

Saravanan R, Ono Y, Isshiki M, Ohno K, Kajitani T (2002a) J Phys Chem Solids 64/1:51–58

Saravanan R, Israel S, Swaminathan S, Kalidoss R, Muruganantham M (2002b) Cryst Res Technol 37:1310–1317

Saravanan R, Ono Y, Ohno IM, Ohno K, Kajitani T (2003) J Phys Chem Solids 64:51–58

Saravanan R, Majella MAA, Jainulabdeen S (2007) Physica B 400:16–21

Saravanan R, Syed Ali KS, Israel S, Pramana (2008) J Phys 70(4):679–696

Shahi K, Wagner JB (1981) Phys Rev B 23(12):6417–6421

Syed Ali KS, Saravanan R, Israel S (2009) J Cryst Growth 311:1110–1116

Takata M, Umeda B, Nishibori E, Sakata M, Saito Y, Ohno M, Shinohara H (1995) Nature 377:46–49

Teeple HO (1954) Nickel and high-nickel alloys. Ind Eng Chem 46:2092–2107

Tobi H, Egami T (1992) Acta Cryst A 48:336–346

Valvoda V (1970) Acta Crystallographica A 26:405–406

Waasmaier D, Kirfel A (1995) Acta Cryst A51:416–431

Weiss RJ (1966) X-ray determination of electron densities. North-Holland Publishing, Amsterdam

Wirojanupatump S, Shipway PH, Cartney DGM (2001) Wear 249:829–837

Wyckoff RWG (1963) Crystal structure, vol 1. Inter-Science publishers, London

Xiangyun Q, Proffen T, Michell JF, Billinge SJL (2005) Phy Rev let 94:177203–177206

Laugier J, Bochu B GRETEP: Domaine universitaire BP 46, 38402 Saint Martin d'Hères. http://www.inpg.fr/LMGP

Saravanan R GRAIN software. http://www.saraxraygroup.net

Wilkinson K. http://www.chemclub.com/teachers

Yamamoto K, Takahashi Y, Ohshima K, Okamura FP, Yukino K (1996) Acta Cryst A52:606–613

Yamamura S, Takata M, Sakata M, Suguwara Y (1998) J Phys Soc Jpn 67:4124–4127

Zwick LZJ, Lugscheider E (2004) Surf Coat Technol 82(1):72–77

Chapter 4
Conclusion

The average and local structures of some elemental metals, alloys and dilute doped alloys have been analysed from X-ray data using versatile techniques like Rietveld method, MEM (maximum entropy method), multipole and PDF (pair distribution function). The materials studied in this research work are listed below.

Metals

- Sodium (Na)
- Vanadium (V)
- Magnesium (Mg)
- Aluminium (Al)
- Titanium (Ti)
- Iron (Fe)
- Nickel (Ni)
- Copper (Cu)
- Zinc (Zn)
- Tin (Sn)
- Tellurium (Te)

Alloys

- Cobalt aluminium (CoAl)
- Nickel aluminium (NiAl)
- Nickel chromium (NiCr)
- Iron nickel (FeNi)

Dilute doped alloys

- Sodium chloride with iron impurities ($Na_{1-x}Ag_xCl$)
- Aluminium, with iron impurities (0.215 and 0.304 wt% Fe)

The materials studied in this research work have been divided into several parts for the exhaustive analytical study and the results of this research work are presented below.

R. Saravanan and M. Prema Rani, *Metal and Alloy Bonding: An Experimental Analysis*, 147
DOI: 10.1007/978-1-4471-2204-3_4, © Springer-Verlag London Limited 2012

4.1 Sodium and Vanadium Metals

- The nature of bonding and the charge distribution in sodium and vanadium metals have been analysed using the reported X-ray data of these metals.
- The bonding in these metals has been elucidated and analysed using MEM (maximum entropy method) and multipole method.
- The nature of bonding is clearly revealed by the two-dimensional MEM maps plotted on (100) and (110) planes and the mid-bond densities are clearly revealed by the one-dimensional electron density along the [100], [110] and [111] direction.
- The mid-bond densities in sodium and vanadium are found to be 0.014 and 0.723 e/Å^3 respectively, giving an indication of the strength of the bonds in these materials.
- In sodium, the very large thermal vibration of the sodium atom is seen in the electron density map on the (100) plane.
- From multipole analysis, the large κ' parameter of the valence part of the sodium atom denotes the contraction of the sodium atom.
- Static and dynamic multipole deformation maps drawn clearly show the thermal effect on the electron density in the core and valence regions.
- The sodium atom is found to contract more than the vanadium atom from the multipole analysis.

4.2 Aluminium, Nickel and Copper

- The average and local structure of simple metals Al, Ni and Cu have been elucidated for the first time using MEM (Maximum Entropy Method), multipole and PDF (Pair Distribution Function).
- The powder X-ray data sets were refined by Rietveld method using the software program JANA 2000.
- The refined structure factors were used for the MEM and multipole analysis.
- The electron density distribution of Al, Ni and Cu has been mapped using the MEM electron density values obtained through refinements.
- The mid-bond density in Ni is found to be larger among the three metal systems along [110] direction. But along the [100] and [111] directions (non-bonding), the element with the higher atomic number has a higher electron density.
- The thermal contribution to the charge density has been realized through the dynamic and static multipole deformation maps.
- From multipole results there is appreciable contraction in Ni compared to the other two systems.

- The observed PDF of Al, Ni and Cu were obtained from the raw intensity data using a software program PDFgetX.
- A comparison between the observed and calculated PDF has been carried out using the software package PDFFIT.
- The PDF results give a clear picture of the local structure in these metals.

4.3 Magnesium, Titanium, Iron, Zinc, Tin and Tellurium

- The average and local structures of magnesium, titanium, iron, zinc, tin and tellurium have been analysed using the MEM (maximum entropy method), and PDF (pair distribution function).
- The structural parameters of the metals were refined from the powder X-ray data with the well-known Rietveld powder profile fitting methodology using the software package JANA 2006.
- The refined structure factors were used for the MEM analysis
- The one-, two- and three-dimensional electron density distributions of Mg, Ti, Fe, Zn, Sn and Te have been mapped using the MEM electron density values obtained through refinements.
- The variation of electron density distribution with respect to the thermal vibration factor is clearly visualised.
- The electron density has been analysed qualitatively and quantitatively.
- The local structure has been analysed for the accurate nearest neighbour distances using PDF (pair distribution function).

4.4 Cobalt Aluminium and Nickel Aluminium Metal Alloys

- The precise electron density distribution and bonding in metal alloys CoAl and NiAl are characterized using MEM (maximum entropy method) and multipole method from reported single-crystal data sets.
- The mid-bond electron densities in these systems are found to be 0.358 and 0.251 e/Å^3 respectively, for CoAl and NiAl in the MEM analysis.
- The two-dimensional maps and one-dimensional electron density profiles have been constructed and analysed.
- The thermal vibration of the individual atoms Co, Ni and Al has also been studied and reported.
- The thermal vibration of Al is more pronounced in CoAl than in NiAl.
- The contraction of atoms in CoAl and expansion of Ni and contraction of Al atom in NiAl is found from the multipole analysis.

4.5 Nickel Chromium ($Ni_{80}Cr_{20}$)

- The alloy $Ni_{80}Cr_{20}$ has been annealed and ball milled to study the effect of thermal and mechanical treatments on the local structure and the electron density distribution.
- The electron density between the atoms was studied by MEM (maximum entropy method) and the local structure using PDF (pair distribution function).
- The electron density is found to be high for ball-milled sample along the bonding direction.
- The particle sizes of the differently treated samples were realized by SEM and through XRD.
- Clear evidence of the effect of ball milling is observed on the local structure and electron densities.

4.6 Silver Doped in NaCl ($Na_{1-x}Ag_xCl$)

- The alkali halide $Na_{1-x}Ag_xCl$, with two different compositions ($x = 0.03$ and 0.10), was studied with regard to the Ag impurities in terms of the bonding and electron density distribution using X-ray single-crystal data sets.
- Electron density distribution and hence the interaction between the atoms is clearly revealed by maximum entropy method (MEM) and multipole analyses.
- The bonding in these systems has been studied using two-dimensional MEM electron density maps on the (100) and (110) planes and one-dimensional electron density profiles along the [100], [110] and [111] directions.
- The mid-bond electron densities between atoms in these systems are found to be 0.175 and 0.183 e/$\mathring{A}^3$, respectively, for $Na_{0.97}Ag_{0.03}Cl$ and $Na_{0.90}Ag_{0.10}Cl$.
- Multipole analysis of the structure has been performed for these two systems, with respect to the expansion/contraction of the ion involved.

4.7 Aluminium Doped with Dilute Amounts of Iron Impurities (0.215 and 0.304 wt% Fe)

- The electronic structure of pure and doped aluminium with dilute amounts of iron impurities (0.215 and 0.304 wt% Fe) has been analysed using reported X-ray data sets and the MEM (maximum entropy method) technique.
- Qualitative as well as quantitative assessment of the electron density distribution in these samples has been made.
- An expansion of size of the host aluminium atom is observed with Fe impurities. Minute changes in the density details and the incorporation of impurities in the host lattice have been revealed.

- The electron density and the mid-bond electron density decrease with more Fe impurity doping as evidenced from the one-dimensional electron densities.
- The bonding strength in Al decreases with Fe addition.

The broad applicability of structure determination and refinement with single crystals and powders make the future of diffraction studies more exciting and essential. The success of extraction of details mainly depends on three parameters, choice of measurement device, pattern profile description and structure solving algorithms. Modern instrumentation, sources yielding data of unprecedented quality and modern analysis methods continue to increase our ability to harvest useful information from the data.

In this research work the charge density of metals and alloys analysed using versatile tools such as maximum entropy method, multipole and pair distribution function have been done for the first time. These analyses give the fine details of the electron environment in the unit cell. The changes in the properties of the alloys due to the inclusion of dopant elements are clearly picturised. These local and average analyses of the materials facilitate the better understanding of their properties and enhance the engineering of new materials.

MIX
Papier aus verantwortungsvollen Quellen
Paper from responsible sources
FSC® C105338

If you have any concerns about our products,
you can contact us on
ProductSafety@springernature.com

In case Publisher is established outside the EU,
the EU authorized representative is:
Springer Nature Customer Service Center GmbH
Europaplatz 3, 69115 Heidelberg, Germany

Printed by Libri Plureos GmbH
in Hamburg, Germany